The Circulatory System

**Other titles in
Human Body Systems**

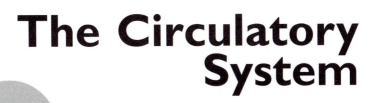

The Circulatory System

Leslie Mertz

HUMAN BODY SYSTEMS
Michael Windelspecht, Series Editor

Greenwood Press
Westport, Connecticut • London

Library of Congress Cataloging-in-Publication Data

Mertz, Leslie A.
    The circulatory system / Leslie Mertz.
        p.   cm.—(Human body systems)
    Includes bibliographical references and index.
    ISBN 0–313–32401–8 (alk. paper)
        1. Cardiovascular system.   I. Title.   II. Human body systems.
    QP101.M4586   2004
    616.1—dc22   2004042449

British Library Cataloguing in Publication Data is available.

Library of Congress Catalog Card Number: 2004042449
ISBN: 0–313–32401–8

First published in 2004

Greenwood Press, 88 Post Road West, Westport, CT 06881
An imprint of Greenwood Publishing Group, Inc.
www.greenwood.com

Printed in the United States of America

The paper used in this book complies with the Permanent Paper Standard issued by the National Information Standards Organization (Z39.48–1984).

10 9 8 7 6 5 4 3 2

Illustrations, unless otherwise credited, are by Sandy Windelspecht.

Contents

Color photos follow p. 116.

Series Foreword

Human Body Systems is a ten-volume series that explores the physiology, history, and diseases of the major organ systems of humans. An organ system is defined as a group of organs that physiologically function together to conduct an activity for the body. In this series we identify ten major functions. These are listed in Table F.1, along with the name of the organ system responsible for the activity. It is sometimes difficult to specifically define an organ system, because many of our organs have dual functions. For example, the liver interacts with both circulatory and digestive systems, the hypothalamus acts as a junction between the nervous and endocrine systems, and the pancreas has both digestive and endocrine secretions. This complex interaction of organs and tissues in the human body is still not completely understood.

This series is unique in that it provides a one-stop reference source for anyone with an interest in the human body. Whereas other references frequently cover one aspect of human biology, from anatomy and physiology to the prevention of diseases, this series takes a more holistic approach. Each volume not only includes a physiological description of how the system works from the cellular level upward, but also a historical summary of how research on the system has changed since the time of the ancients. This is an important aspect of the series, and one that is frequently overlooked in modern textbooks. In order to understand the successes and problems of modern medicine, it is first important to recognize not only the achievements of the past but also the misunderstandings and challenges of the pioneers in medical research.

For example, a visit to any major educational institution reveals large lecture halls, where science instructors present material to the students on the

TABLE F.1. Organ Systems of the Human Body

Organ System	General Function	Examples
Circulatory	Movement of chemicals through the body	Heart
Digestive	Supply of nutrients to the body	Stomach, small intestine
Endocrine	Maintenance of internal environmental conditions	Thyroid
Lymphatic	Immune system, transport, return of fluids	Spleen
Muscular	Movement	Cardiac muscle, skeletal muscle
Nervous	Processing of incoming stimuli and coordination of activity	Brain, spinal cord
Reproductive	Production of offspring	Testes, ovaries
Respiratory	Gas exchange	Lungs
Skeletal	Support, storage of nutrients	Bones, ligaments
Urinary	Removal of waste products	Bladder, kidneys

anatomy and physiology of the human body. Sometimes these classes include laboratory sessions, but in the study of human biology, especially for students who are not bound for professional schools in medicine, the student's exposure to human biology typically centers on a two-dimensional graphic. Most educators accept this process as a necessary evil of the educational system, but few recognize that, in fact, the large lecture classroom is the product of a change in Egyptian religious beliefs before the start of the current era. During the decline of the Egyptian empires, and the simultaneous rise of the ancient Greek culture, the Egyptian religious organizations began to forbid the dissection of the human body. This had a twofold influence on medicine. First, the ending of human dissections meant that medical professionals required lectures from educators, instead of participation in laboratory-based education, which led to the birth of the lecture hall. The second consequence would plague modern medicine for a thousand years. Stripped of their access to human cadavers, researchers studied other "lesser" animals, and extrapolated their findings to humans. The practices of the ancient Greeks were passed on over the ages and became the basis for the study of modern medicine. These traditions continue to this day throughout the educational institutions of the world.

The history of human biology parallels the development of modern science. In the seventeenth century, William Harvey's study of blood circulation challenged the long-standing belief of the ancient Greeks that blood was produced in the liver and consumed in the tissues of the body. Harvey's pioneering experimental work had a strong influence on others, and within a century the legacy of the ancient Greeks had collapsed. In the eighteenth century a group of chemists who focused on the chemical reactions of the human body, called the iatrochemists, began to apply chemical laws to human physiology. They were joined by the iatrophysicists, who believed that the human body must operate under the physical laws of the universe. This in turn led to the beginnings of organic chemistry and biochemistry in the nineteenth century, as scientists focused on identifying the building blocks of living cells and the chemical reactions that they utilize in their metabolism.

In the past century, especially in the last three decades, the rapid advances in technology and scientific discovery have tended to separate most sciences from the general public. Yet despite an ongoing trend to leave the majority of the physical sciences to the scientists, interest in human biology has actually increased among the general population. This is primarily due to medical discoveries that increase not only lifespan but also healthspan, or the number of years that people live disease free. But another important aspect of this trend is the desire among the general public to be able to ask intelligent questions of their physicians and seek additional information on prescribed medications or procedures. In many cases this information serves as a system of checks and balances on the medical profession, ensuring that the patient is kept well informed and aware of the fundamentals regarding the procedure.

This is one of the most remarkable ages in the study of human biology. The recently announced completion of the Human Genome Project is an indication of how far biology has progressed. Barely fifty years ago, scientists were first discovering the structure of DNA. They now are in possession of an entire encyclopedia of human genetic information, and although they are not yet exactly sure what the content reveals, scarcely a week goes by without a researcher announcing a medical discovery that was made possible by the availability of the complete human genetic sequence. Coupled to this are the advances in the development of pharmaceuticals and treatments that were unheard of less than a decade ago.

But these benefits to society do not come without a cost. The terms stem cells, cloning, and gene therapy no longer belong to the realm of science fiction. They represent advances in the sciences that may hold the key to increased longevity. However, in many cases they also produce ethical and moral questions of society: Where do medical researchers obtain the embryonic stem cells for their work? Who will determine if humans can be cloned? What are the risks of transgenic organisms produced by gene therapy? These are just a few of the potential conflicts that face modern society. Only a well-educated general public can intelligently survey the pros and cons of an ethical or moral decision regarding medical science. Armed with information, concerned people can participate in the democratic process of informing their elected officials of their concerns. Science education is an important aspect of citizenship, and thus the need for series such as this to present information to the general public.

This volume covers the biology of the circulatory system. From an educational perspective, the circulatory system represents one of the most-studied organ systems in the human body. The route of the blood in the body, as well as the structure and operation of the heart, are the focus of many introductory discussions of the human body. Yet the circulatory system is much more complex than a simple series of tubes. The circulatory system represents not only a conduit for the movement of blood, but also for the majority of nutrients, hormones, and waste materials in the body. This system is not just a series of tubes, but rather a highly regulated series of pathways that effectively move fluids to parts of the body as needed. As a component of the cardiovascular system, effective circulation is vital for the chances of a long, healthy lifespan. Circulation problems such as hypertension or atherosclerosis are major health concerns in the Western world and are becoming more common in young people. For this reason, an understanding of the circulatory system is vital for students of science at all ages.

The ten volumes of the *Human Body Systems* are written by professional authors who specialize in the presentation of complex scientific ideas to the general public. Although any book on the human body must include the terminology and jargon of the profession, the authors of this series keep it to a minimum and strive to explain the concepts clearly and concisely. The series is ideal for the public libraries, as well as for secondary school and introductory college libraries. In addition, medical professionals or anyone with an interest in human biology would find this series a useful addition to their personal library.

Michael Windelspecht
Blowing Rock, North Carolina

Acknowledgments

Sincere thanks go to Stephen Zaglaniczny, who was a constant source of support over the many months of work that went into this project, and to the wonderfully helpful librarians at public and university libraries throughout the state of Michigan.

Introduction

Animals possess an array of internal circulatory systems for transporting materials through the blood and to every part of the body. Depending on the type of animal, the system may be quite simple or very complex. Earthworms, for example, have a series of "hearts" that are little more than pulsating blood vessels to assist the transit of blood through other vessels and to body organs. Insects and some other invertebrate animals have what is known as an open circulatory system that forgoes the network of vessels and instead usually delivers blood through one long, dorsal vessel that empties into a large cavity, or sinus. In that flooded cavity, the body organs are actually bathed in blood. As organisms become more complex and larger, a closed circulatory system is the norm. In a closed system, the blood makes its route through the body and to the tissues in vessels. Earthworms have a closed circulatory system, and so do all vertebrates, including humans.

The purpose of this volume is to examine the structure and function of the human circulatory system, which is also known as the cardiovascular system because it includes the heart, or *cardium* from the Greek word for heart, and the blood vessels, or *vasculature*. The blood and all of its cells are also part of this system. The circulatory system serves as the body's delivery method, picking up oxygen from the lungs and dropping it off at tissue and organ cells around the body, then gathering carbon dioxide from the tissues and organs, and shipping it off to the lungs. It also distributes nutrients from the digestive system, transports chemical messages from the brain and other organs to various sites in the body, and provides the route and means for the body to mount a defense against bacterial infection. It even maintains the internal temperature by shuttling excess heat from the core of the body to the outside.

The force behind the system is a fist-sized organ weighing less than a pound. This organ, the heart, usually contracts more than once a second. It pumps blood into a network of vessels that become increasingly branched and reach every cell in the human body. Once there, blood cells and tissue cells exchange oxygen, carbon dioxide, nutrients, and an assortment of other materials. The blood then collects in small vessels that merge with larger and larger vessels on their return path to the heart. There, the cycle repeats. This unending loop is the basis for continued existence. If any part of the system breaks down, life is jeopardized. Yet, the circulatory system normally is amazingly reliable. It continues working second by second, minute by minute, and hour by hour, consistently transporting the essential gases and materials to sustain a human being for many, many years.

This book is part of a ten-volume reference series on the human body. Designed not only for science and medical students, but also for anyone interested in the human heart and circulation, this book provides a summary of the parts of the cardiovascular system and their functions; the history of discovery, as well as new research; and an overview of medical conditions associated with the heart, blood vessels, and blood.

A list of interesting facts about the remarkable human cardiovascular system follows this chapter. The anatomy and physiology of the cardiovascular system are covered in the first section of the book (Chapters 1–7). This includes an introduction to the parts of the heart, the makeup of the blood vessels, and the variety of blood cells. The section also explains how they work—separately and together—to maintain life in the intriguingly complex human body.

The second section (Chapters 8–9) demonstrates the sometimes-tortuous and sometimes-accelerated path to current understanding of the cardiovascular system. The chapters point out many of the crucial discoveries that advanced knowledge, along with some of the hypotheses that stalled scientific progress for decades or, in some cases, for centuries.

The third section of the book (Chapters 10–11) delves into the wide range of medical conditions associated with the cardiovascular system. From blood diseases to heart attacks, and from genetic disorders to accident-related blood loss, Chapter 10 includes a definition of each medical condition, its symptoms and causes, and a variety of treatment options. Small articles throughout the chapter target topics of interest, including new findings and areas of study. Chapter 11 examines current thinking on the importance of diet, exercise, and other lifestyle choices to cardiovascular health.

The back of the book contains a list of commonly used acronyms, a glossary, a compilation of online resources, a bibliography, and a full index. Because the book is written for a wide audience encompassing science and medical students, as well as others interested in the human heart and circulation, readers will find that all key terms are printed in **bold** type at their first mention, defined in the text, and listed again in the comprehensive glossary. The organization of this volume and series makes this work attractive for secondary-school libraries, and undergraduate higher-education colleges and universities where students may be seeking general information on the cardiovascular system. Community libraries in search of a comprehensive general reference volume on the cardiovascular system, as well as individuals with an interest in science, history, or medicine, will also find this work a useful addition to their collections.

INTERESTING FACTS

▶ At any given time, the veins and venules typically hold about two-thirds of the blood flowing through the body.

▶ As the heart contracts and blood rushes into the aorta, it is traveling at a speed of about 8 inches (20 centimeters) per second.

▶ Even in a person who's resting, blood issuing from the heart can travel down to the person's toes and back to the heart in just a minute. When a person is exercising heavily, that trip can take just 10 seconds. On average, every red blood cell completes the heart-to-body-to-lungs circuit 40–50 times an hour.

▶ If all of the blood vessels in an average adult were strung together end to end, they would reach at least 60,000 miles long, more than twice the distance around the Earth's equator. The capillaries alone make up 60 percent of that total.

▶ Every second, 10 million red blood cells die in the normal adult. The body replaces them just as quickly, however, so the total number remains constant.

▶ In the average adult, the heart weighs less than three-quarters of a pound—about 11 ounces (310 grams). In any given person, it's about the size of his or her fist.

▶ The heart beats an average of 72 times a minute with a typical at-rest volume of 75 ml of blood pumped with each beat. Using those figures, a 75-year-old's heart has contracted more than 2.8 billion times and pumped more than 212 million liters of blood in his or her lifetime.

▶ When a person is resting, the left ventricle pumps about 4–7 liters of blood every minute. In a well-trained athlete who is doing strenuous exercise, that amount can rise to almost 30 liters per minute.

▶ Heart rate changes greatly during child development. The typical heart rate in a newborn is 130 beats per minute (bpm). It drops to 100 bpm by the time the child reaches 3 years old, 90 at 8 years old, and 85 at 12 years old.

▶ In an increasingly common practice, people are donating blood for use in their own upcoming surgeries. Called autologous blood donation, it helps patients ensure safe transfusions.

▶ Among Americans overall, the most common blood type is almost tied between O-positive and A-positive, with 38 percent of the population having O-positive and 34 percent having A-positive. The three least common blood types are AB-positive, 3 percent; B-negative, 2 percent; and AB-negative, 1 percent. Within specific ethnic groups, those numbers change. For example, among African Americans, 47 percent have O-positive blood and 24 percent have A-positive.

▶ If the blood supply to the brain is stopped, a person can remain conscious for about nine seconds.

The Human Circulatory System:
A Living River

An amazing river system flows through the human body. After a person takes a breath of air, the oxygen is swept into the current and rushes to muscles, the brain, or another part of the body. Shortly after a Sunday dinner, nutrients begin to make their way into this same system for dispersal throughout the body. Alongside them, bacteria-fighting cells race to the site of an infection.

This remarkable riverway is the **circulatory system**: the **heart**, **blood vessels**, and **blood**. In a river system, water flows with a current between banks. A typical river system comprises tiny creeks, usually with a very slow current; larger, faster-moving river branches; and the main river with its strong flow. Likewise, the circulatory system has a liquid that flows with a current within a confined space. Blood replaces water. Instead of a river bed, the blood flows inside of tubes, the blood vessels. Here and there, smaller branches separate from or connect into the main bloodstream. Blood from these branches eventually flows into or collects from even tinier vessels, which have a much slower current than the main bloodstream.

Unlike a riverway, however, the circulatory system is—as its name implies—clearly circular. Rivers generally empty at some point into an ocean, lake, or reservoir. The bloodstream flows up to the head, down to the toes, and out to the fingers, but for the most part it stays inside the body and within the blood vessels. A bit of blood that begins in the heart can travel anywhere in the body, but it eventually returns to the heart for its next trip out to the body. In addition, the circulatory system requires a central power unit to

In some regards, water flows through a river system much like blood moves through the human circulatory system. © Photodisc.

maintain its current. Water that races down the side of a mountain can rely on gravity, but such a dependence in a human being would simply pool the blood in whatever part of the body was closest to the ground. Instead, the circulatory system depends on the heart, a powerful, continuously running pump, to give the blood an initial push as it begins its circuit through the body.

A riverway can transport all sorts of things along its route. The human circulation is a transportation system, too. Oxygen from the lungs, sources of nutrition from the digestive system, antibodies to fight disease, and other substances all enter the bloodstream for shipment to different parts of the body. Even waste products join the bloodstream as the first leg of their journey for eventual transport out of the body.

The circulatory system, however, does much more than ship materials to and fro. It helps maintain the body temperature, so a person doesn't overheat or overcool. For example, when a person exercises and begins to warm up, the blood transports some of that heat from the body's core out to the skin surface where heat can dissipate. A variety of other mechanisms operate, as needed, to regulate body temperature. The circulatory system is so well adapted to this task that a temperature change of only a few degrees from the typical 98.6°F can signal a serious illness.

Although the brain communicates to the rest of the body via the nervous system, it also communicates through chemical compounds called **hormones**. Other organs and tissues similarly produce and release hormones, most of which move through the body by way of the circulatory system. These chemical messengers play a variety of important roles during a person's life. For example, the **pituitary gland** in the brain secretes hormones that stimulate a mother to produce milk to nurse her child, or trigger some of the changes necessary for the sexual development of children into adolescents and adolescents into adults. The **adrenal gland**, which lies atop the kidney, releases **adrenaline** (also called epinephrine), the hormone responsible for the quickened heartbeat, rapid breathing, goosebumps, and hair that "stands on end" when a person is startled. Adrenaline causes many other so-called fight-or-flight responses when a person is faced with different forms of stress.

Overall, the human circulatory system is vital to nearly every physiological function in the body. Organs and cellular tissue cannot survive without oxygen and nutrients, both of which are delivered by the bloodstream. A person's internal body temperature would fluctuate wildly without the moderating effect of the circulatory system. If hormones could no longer pass into the circulatory current, the brain and many other organs would lose a significant method for communicating with the rest of the body. The circulatory system is as vital to human life as it is fascinating.

The following sections provide an introduction to the three main components of the circulatory system:

- The heart
- The blood vessels
- The blood

THE HEART

The heart is a hollow, muscular organ about the size of a softball: small enough at about 10 ounces (280 grams) to fit between the breast bone and the backbone, but powerful enough to continuously force blood out into all of the blood vessels during a person's entire natural lifespan. That is quite a feat, considering that a person's heart beats an average of 72 times per minute—more than once every second—which adds up to 103,680 times each day and nearly 38 million times per year. With every beat, the heart pumps 2–3 ounces (60–90 milliliters) of blood, an amount equivalent to about two good gulps of soda. If the blood were not contained within the circulatory system, the power of that single pump would shoot a spurt of blood about 6.6 feet (2 meters) into the air. Instead, the energy of each heartbeat is directed to moving blood into and through the blood vessels.

The heart can be described as having two distinct halves. The right half forces blood to the lungs, where it picks up the oxygen that the lungs draw in every time a person takes a breath. From the lungs, the newly oxygenated blood enters the left half of the heart, which pumps the blood out to the rest of the body. Once the blood completes its route through the body, it returns to the right half of the heart, where the whole cycle repeats.

Throughout this cycle, the blood all flows in the same direction: right heart to lungs to left heart to body and back to the right heart. To ensure that the heart only pumps in one direction, the heart has a series of valves that only open one way. These **heart valves** swing shut if blood starts moving the wrong way, thus preventing backflow and preserving the forward movement of the bloodstream. When a physician uses a stethoscope to check a person's heartbeat, he or she is actually listening to the heart valves. For each heartbeat, two sounds are audible: "lubb-dupp." The "dupp" is the sound of certain heart valves closing. The "lubb" is associated in part with

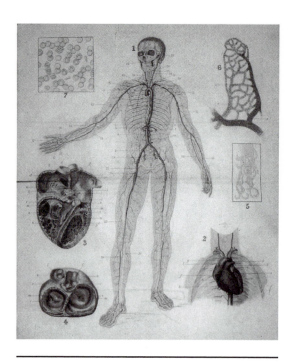

The heart and arteries. Engraving from *Atlas of Human Anatomy and Physiology*, London, 1857. © National Library of Medicine.

the contraction of the heart as it pumps blood. By pressing the fingers gently on either side of the front portion of the neck, a person can feel his or her pulse, which is the blood rushing away from the heart in response to its pumping force. In other words, each thump of this neck pulse, which is named the **carotid pulse**, is tied to the "lubb" sound.

Overall, the heart's primary function is to move blood and to keep moving it as long as a person lives. The heart pumps blood all day and all night, during strenuous exercise and during sleep, usually without so much as a thought from the person in whose chest it resides.

THE BLOOD VESSELS

Once the blood leaves the heart, it uses the blood vessels to travel to the organs, tissues, and cells throughout the body (see color insert). Also known as the **vasculature**, blood vessels have the job of transporting blood. The circulatory system has five main types of blood vessels:

- Arteries
- Arterioles
- Capillaries
- Venules
- Veins

The vast majority of the **arteries** and **arterioles** deliver oxygen-rich blood from the heart to the body cells. The arteries are typically the larger of the two. Using the riverway analogy, the arteries would compare to the main river body. Each major part of the body has an artery that feeds it. These arteries divide into smaller vessels, the arterioles, which branch into even tinier vessels.

These tiny vessels—the creeks of the riverway—are the **capillaries**, which bring blood to each of the 100 trillion cells in the human body. In fact, the vast majority of cells are so close to the nearest capillary that even a single hairbrush bristle couldn't fit between them. Capillaries are basically exchange sites, where the blood drops off oxygen, food, and other materials, and acquires such waste products as carbon dioxide or **urea**. Carbon dioxide builds up in cells as each cell uses, or metabolizes, food. The blood picks up this gas, which eventually leaves the body as a person exhales. Likewise, as a cell metabolizes proteins, urea forms. A solid rather than a gas, this waste product rides in the bloodstream and ultimately winds up in the urine.

After the exchange between cells and capillaries, the blood begins its journey back to the heart. Capillaries flow into small blood vessels known as **venules**. The venules join together and eventually merge into larger vessels, the **veins**, which are found in every major region of the body. Blood moves through the veins until it returns to the heart to begin another round trip.

THE BLOOD

Every person's body holds about about 1.3 gallons (5 liters) of blood. Although it may appear to be just a thick, red fluid, blood is actually filled with many kinds of **blood cells**, all of them floating in a clear liquid called **plasma**. Blood gets its overall red coloration from the numerous red cells—logically called **red blood cells** or **erythrocytes** (*erythros* is Greek for red and *cyt* comes from the Greek *kytos* for cell)—that lie within. These red blood cells are responsible for gathering and delivering oxygen from the lungs to the body. A drop of blood (one cubic millimeter) from the average person contains 4.5–5.5 million red blood cells. That adds up to approximately 35 trillion red blood cells in the typical adult's circulatory system.

The blood also carries other types of cells, including **white blood cells** and **platelets**. White blood cells are also called **leukocytes** (*leukos* is Greek for white). These cells form a defensive barrier by coursing through the body and gobbling up bacteria that might otherwise be dangerous. Most people are familiar with white blood cells by their less dainty name: pus. After their encounters with bacteria, white blood cells typically die. Pus is the accumulation of dead white blood cells that forms at the site of an infection. Leukocytes have additional functions, but they are primarily the body's defenders. Platelets, on the other hand, can be described as battlefield medics. The job of these round or oblong disks is to rush to the site of a wound, stick together to bridge the injury, and create a **blood clot**—a quick fix to what might otherwise become a serious loss of blood.

Besides the red blood cells, white blood cells, and platelets, the blood still has room to carry a wide range of materials to and from the body's cells. **Antibodies** ride the bloodstream to arrive at infection locations, where they use their own unique tactics to fight off bacteria or other invaders. The blood transports **electrolytes**, which are charged particles that have a variety of functions in cells. Examples are calcium (Ca^{2+}) and magnesium (Mg^{2+}), which stabilize cellular membranes. Food, including vitamins, migrates from the digestive system to the cells through the bloodstream. Hormones and even water employ the circulatory system for transit.

As a whole, the circulatory system is very much like a riverway. Both a river and the bloodstream have a liquid that moves with a current along a set route. Both carry various materials to new destinations. One can play a role in a healthy ecosystem, while the other is vital to the health, and indeed the life, of a single human being. The heart, blood vessels, and blood are indeed part of a remarkable internal "riverway" that is essential to nearly every physiological function within the human body. The following chapters will provide a closer look at this amazing system.

The Blood: A Vital Mixture

Blood is more than just a simple, red liquid. It is actually a clear, somewhat gold-colored, protein-rich fluid crowded with red and white cells. The preponderance of red cells gives it the scarlet cast. When separated from the rest of the blood, the clear fluid, called plasma, has a more watery consistency. The reason that blood is more like syrup than water is the addition of red and white cells, and platelets, which combine to make up 40–45 percent of blood volume. Just as a glass of mud is more difficult to pour than a glass of water, because mud is actually a mixture of water plus dirt particles, blood is thicker because its plasma is laden with red and white blood cells. From this standpoint, blood truly is thicker than water. The "thickness" of a liquid is known as its **viscosity**. The slower something flows, the more **viscous** it is. Blood, for example, is three to four times more viscous than water.

All of the various components of blood have vital functions. As an example, the plasma serves as the liquid that suspends the red and white blood cells, along with all of the other chemical compounds and various materials that use the bloodstream to travel throughout the body. It also regulates the movement of heat from the body's core to the skin, the head, and the extremities. The red blood cells have a primary role of transporting oxygen from lungs to cells, while the white blood cells help defend against infection from invading organisms and foreign proteins. Table 2.1 describes some of these components in greater detail.

RED BLOOD CELLS

Of the 5.5 quarts (5.2 liters) of blood in an average person, the red blood cells are, by far, the most prominent cellular component. Red blood cells,

TABLE 2.1. Components of the Blood

	Size	Lifespan	Number
Red blood cells* (Erythrocytes)	6.5-8.8 μm	120 - 180 days	5.5 x 10^{12}/L in males 4.8 x 10^{12}/L in females
White blood cells (Leukocytes)	7-18 μm	variable	4 -11 x 10^9/L
Platelets	about 3 μm	4-10 days	150 - 400 x 10^9/L

*The number of red blood cells increases among persons living at high altitudes. In extreme altitudes, individuals may have 50 percent more red blood cells than the amounts shown here.

or erythrocytes, number about 28.6 trillion in the average male and 24.8 trillion in the average female. It follows, then, that red blood cells are microscopic. Ranging in size from 2.6–3.5×10^{-4} inches (6.5–8.8 μm), red blood cells are disk-shaped cells with concave depressions in the centers of both sides (see color insert). If red blood cells were placed side by side in a line, it would take between 2,900 and 3,900 cells before the line would reach about an inch (2.5 cm) long (see color insert). Red blood cells must be flexible, too. The flexibility is critical, because they have to bend, twist, and otherwise deform to squeeze through the tiny capillaries that serve as gateways to the tissue. Another identifying feature of red blood cells is the lack of nuclei, a characteristic that sets them apart from other blood cells.

The primary duty of red blood cells is to transport oxygen and carbon dioxide. After a human breathes in oxygen, the red blood cells deliver it to the tissues. As tissue cells use the oxygen, carbon dioxide begins to accumulate. The red blood cells then pick up the carbon dioxide waste product and transport it back to the lungs, where it is discharged during exhalation.

The jobs of picking up and delivering oxygen and carbon dioxide are accomplished through a large chemical compound known as **hemoglobin** (see Figure 2.1). Located within red blood cells, hemoglobin also gives red blood cells their red color. The more oxygen the hemoglobin is carrying, the brighter red the blood. When the blood is carrying carbon dioxide rather than oxygen—when deoxygenated blood is returning from the tissues back to the lungs—blood takes on a dark maroon hue. The change in color is actually a result of a slight change in the three-dimensional configuration of hemoglobin when it is carrying oxygen. Hemoglobin itself is a combination of a simple protein and an organic structure that contains iron ions. The protein, called a globin protein, contains four polypeptide chains, which are short strings of **amino acids**, the building blocks of proteins. In adults, the four chains normally come in two varieties: a pair of alpha chains and a pair of beta chains. Such hemoglobin is designated hemoglobin A. (Fetal hemoglobin is slightly different with two gamma chains replacing the two beta polypeptide chains seen in adults. The fetal gamma chain has an altered ability to carry oxygen. See Chapter 6 for additional information.) Each polypeptide chain is coupled with a separate iron ion, which is bound in a ringlike chemical structure known as a **heme group**. This heme group is the part of hemoglobin that actually binds oxygen for transport through the bloodstream.

Each red blood cell contains approximately 300 million molecules of hemoglobin, and every one of those units can bind with a total

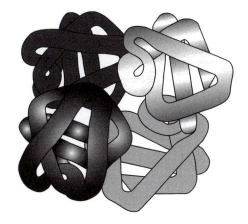

hemoglobin

Figure 2.1. The hemoglobin molecule in a normal adult.
Hemoglobin is made up of four globin chains: two beta and two alpha chains. Within each globin chain sits one iron-containing heme group. The heme group binds oxygen for transport from the lungs to the body tissues.

of four oxygen molecules—one oxygen for each of the four heme groups. Hemoglobin doesn't always bind oxygen, however. It has a differential ability to bind oxygen, which means that it picks up oxygen when the oxygen content in surrounding tissues is high, as it is in the lungs, and drops off oxygen when the oxygen content in the surrounding tissues is low, as it is in the tissues. This relative oxygen concentration is known as **partial pressure**. This property makes hemoglobin an ideal oxygen transportation vehicle. The high partial pressure existing in the lungs stimulates hemoglobin to load up with oxygen, and the low partial pressure in tissues triggers hemoglobin to release it. Sometimes, particularly under conditions of high acidity in the blood or elevated temperature, the red blood cells' affinity for oxygen can drop. This can cause oxygen delivery to tissues to similarly drop. (The relationship between pH and oxygen is discussed further in the section on plasma.)

The blood is also an excellent carrier for carbon dioxide, a byproduct of cell metabolism. As carbon dioxide enters the red blood cells at a tissue site, it lowers the hemoglobin's affinity for oxygen, which further facilitates the discharge of oxygen into the tissues. Once in the blood, the carbon dioxide mostly travels either bound to hemoglobin or as bicarbonate ions (HCO_3^-) that form when carbon dioxide is hydrated (combined with water). The majority of CO_2 takes the latter form. Once the blood arrives at the lung, the bicarbonate ions revert to their original CO_2 state and depart through the lungs. The hemoglobin drops off its carbon dioxide for a similar exit, and the hemoglobin is ready to accept oxygen once again.

Besides oxygen, hemoglobin can lock onto dangerous gases like carbon monoxide (CO), a molecule that contains only one oxygen molecule, *monoxide* (carbon *di*oxide has two). Carbon monoxide is hazardous to human health because hemoglobin binds with carbon monoxide molecules 200 times more readily than it does with oxygen molecules. Therefore, it preferentially binds carbon monoxide instead of oxygen, which can severely reduce oxygen flow to tissues and can quickly become fatal. Cigarette smoke, as well as emissions from automobiles and many home heating systems, contains carbon monoxide. The potential for **carbon monoxide poisoning** from this gas, which is colorless and odorless, has prompted health professionals to warn people against running a car in a closed garage and to recommend the use of carbon monoxide detectors to check for a buildup of the gas in a heated home. (See the section on carbon monoxide poisoning in Chapter 10.)

WHITE BLOOD CELLS

White blood cells have a completely different function than red blood cells. White blood cells, or leukocytes, are part of the body's defense team and can actually move out of the bloodstream to do their work in the tissues (see Figure 2.2). Adults may have anywhere from 20 to 60 billion white cells in the bloodstream, far fewer than the 25 trillion red blood cells in the human body. White blood cells come in three main types:

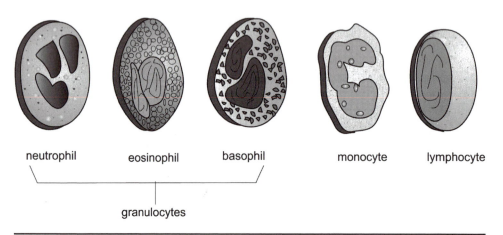

neutrophil   eosinophil   basophil   monocyte   lymphocyte

granulocytes

Figure 2.2. White blood cells come in three main types: granulocytes, monocytes, and lymphocytes.
The granulocytes, so called because of their grainy appearance, include the neutrophils, eosinophils, and basophils. White blood cells are part of the body's defense system against foreign materials and invading microorganisms.

- Granulocytes, including neutrophils, eosinophils, and basophils
- Monocytes
- Lymphocytes

Granulocytes are the most abundant type of white blood cell, comprising seven out of every ten leukocytes. They are named granulocytes based on the grainy appearance of the cytoplasm, the part of the cell outside the nucleus but inside the membrane. All granulocytes also have distinctive lobed nuclei. Depending on which type of biochemical dye best stains them, granulocytes are further subdivided into **neutrophils**, **eosinophils**, or **basophils**. Neutrophils stain with neutral dyes, basophils stain with basic dyes, and eosinophils readily stain with the acid dye called eosin.

In addition to their different staining properties, the three types of granulocytes have separate functions. Neutrophils, the most common granulocyte with up to about 5.2 billion cells per quart (5 billion cells per liter) of blood, engulf and destroy small invading organisms and materials, which are collectively known as **antigens**. Averaging about twice the size of red blood cells, neutrophils are a main bodily defense mechanism against infection and are particularly suited to consuming bacteria. This process of engulfing and destroying bacteria and other antigens is called **phagocytosis** (see Figure 2.3). The sequence of events begins with the human body recognizing that a foreign material has invaded. Antigens are different from the body's own cells and trigger the body to enter its defense mode. If the infected site is within the bloodstream, the neutrophils remain there, but if the site is in the tissues, the neutrophils will flow out through the capillaries to flood the area of infection directly. Each neutrophil at the infected site stretches a bit of its tissue, called a pseudopod, toward and then around the invader. Once the invader is contained inside the neutrophil, an organelle called a lysosome finishes the job by using its internal battery of enzymes and hydrogen peroxide to digest the material. Usually, the neutrophils are able to kill the bacteria quickly, but sometimes the toxins in the bacteria are fatal to the neutrophil. The result is pus, a mixture of mostly dead bacteria and leukocytes that perished during the battle.

Basophils, which are the smallest and least common type of white blood cells, appear to be active in the inflammatory process. While basophilic activity is not fully understood, scientists do know that basophils release substances like histamine and serotonin. Histamine helps maintain a free flow of blood to inflamed tissues, particularly by dilating blood vessels. Serotonin is similarly vasoactive. Basophil granules also contain heparin, which helps prevent blood from clotting. See the volume on the Lymphatic System in this series.

The third group of granulocytes, the eosinophils, likewise demand additional study. Evidence indicates that they engage in minimal phagocytosis.

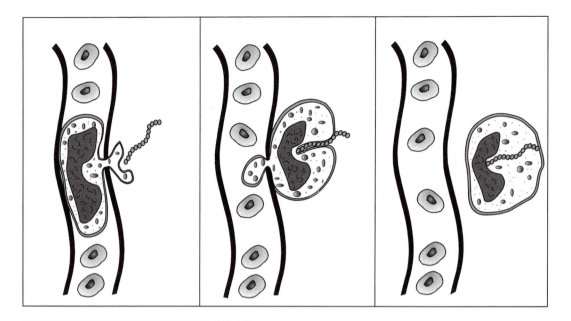

Figure 2.3. Phagocytosis.
Leukocytes are primary players in the body's defense mechanism. Here, a leukocyte protrudes from a blood vessel, and surrounds an invading bacterium. Once the bacterium is engulfed, it is destroyed. This process of engulfing and destroying materials is called phagocytosis.

Their primary tasks are to moderate allergic responses and to help destroy parasites. They accomplish the latter by using unique proteins that are toxic to specific invading organisms. Although they are not quite as motile as the basophils and neutrophils, eosinophils do slowly travel through the body.

The second major type of white blood cell is the **monocyte**, the largest cell in the bloodstream. Monocytes are much larger than red blood cells, and have diameters of $3.9–11.8 \times 10^{-4}$ inches (10–30 μm). They are, however, only temporary residents in the bloodstream, remaining in the blood for about three days before moving into tissues where they become macrophages, large cells that engage in phagocytosis. Just as the neutrophils do, the highly motile monocytes latch onto invading organisms, then literally devour them with a mixture of highly reactive molecules. The neutrophils typically target smaller organisms, like bacteria, while the macrophages take on larger invaders, even **protozoa**, and also remove old cells and other detritus from the bloodstream.

Lymphocytes, the third major type of leukocyte, are the second most common white blood cell, but they frequently aren't in the bloodstream either. They usually reside in the **lymph**, a clear, yellowish fluid that exists around and between cells in the body tissues. This fluid, which is about 95 percent water, enters the bloodstream mainly through one of two ducts, and carries lymphocytes into the blood with it. Lymphocytes come in two main varieties: **B lymphocytes**, or **B cells**, and **T lymphocytes**, or **T cells**. Both B and T cells have antigen receptors on their cell surfaces. These receptors are highly specific. In other words, one particular form of lymphocyte can only bind to one type of foreign material, much as different keys fit different locks. The specific area of an antigen to which the B cell receptor binds is called an **epitope**. For more information on the functions of B cells and T cells, see the Lymphatic System volume in this series.

PLASMA

The circulatory system reaches just about everywhere in the human body, so the volume of plasma is fairly high. In fact, plasma makes up about 5 percent of a normal human's body weight. Plasma itself is a solution of about 90 percent water, 7–9 percent proteins, and roughly 1 percent ions, which are either positively or negatively charged molecules of such chemicals as sodium, calcium, and potassium. The remainder includes dissolved organic nutrients, gases, and waste products. Some textbooks refer to the plasma as the extracellular matrix of blood. In other words, it is the portion of blood that lies outside the red and white blood cells, and that provides the blood's liquid "structure."

Plasma is mostly water, but it is the smaller ion and protein components that draw biologists' interest. Their roles include regulating the blood's volume and viscosity; maintaining the steady **pH level** required by the muscular, nervous, and other major physiological systems; facilitating the transport of various materials; and assisting in bodily defense mechanisms, including the immune response and blood clotting.

Plasma proteins consist of **albumins** (the most abundant plasma protein), **globulins**, and **fibrinogen**, most of which are synthesized in the liver. The ions in plasma include both positively and negatively charged varieties. Sodium (Na^+), calcium (Ca^{2+}), potassium (K^+), and magnesium (Mg^{2+}) carry positive charges and are called **cations** (pronounced "cat-ions"). **Anions** ("ann-ions") are negatively charged, and in the plasma include chloride (Cl^-), bicarbonate (HCO_3^-), phosphate (HPO_4^{2-} and $H_2PO_4^-$), and sulfate (SO_4^{2-}). Most of the ions in the plasma of humans and other mammals are sodium and chloride. The general public is familiar with the two ions in their combined form of NaCl, or normal table salt.

Because blood vessels are permeable to water, water can move freely into and out of the blood. The more water in the blood, the greater the overall blood volume. A change in the concentration of different ions or proteins can play havoc with the amount of water in the blood, and too much or too little water can have damaging effects on a person's health. The body's ion and protein concentrations keep the blood volume from plummeting too low or rising too high. Plasma proteins and ions make up 10 percent or less of the plasma volume, but they are critical in regulating how watery the plasma, and thus the blood, is. Sodium chloride (NaCl) and sodium bicarbonate (NaHCO$_3$) are the key ion regulators of the amount of water in the plasma, while the albumins are the primary proteins involved in the water content of the plasma. This system is based on a balance between the liquid inside the blood vessels and the liquid outside. **Osmosis** is a process that seeks to equalize the water-to-solute ratio on each side of a water-permeable membrane. In other words, if the water on one side of the membrane has twice as many dissolved materials, called **solutes**, additional water will move by osmosis across the membrane and into the side with the higher solute concentration. This action adjusts the solute concentration so that the ratio of water to solutes on each side of the membrane is the same. In blood vessels, water likewise moves in and out based on the relative solute concentrations within and beyond the vessels. As it turns out, the water content in the circulatory system sometimes falls too low, because the sheer force of blood rushing away from the heart literally pushes water out of the smallest blood vessels, the capillaries. The system remains stable because plasma proteins and the albumin proteins, in particular, diffuse very poorly through capillary walls and therefore allow the osmotic pressure gradient to draw water back into the capillaries.

Osmosis alone isn't enough to maintain the proper solute concentration in blood. Blood carries all kinds of solutes, including organic molecules like food, cholesterol and other fats, waste products, and hormones, yet the blood isn't continually flooded with water. The reason is that ions, specifically sodium, use pumps that reside in the membranes of the cells within the blood vessel to actively drive sodium molecules out of the vessels, leaving behind a lower concentration of solutes and, in turn, requiring less water to enter by osmosis. This balance between osmosis and ion pumps is vital to regulating blood volume.

Many bodily systems also depend on a certain pH level, which is a measure of acidity or alkalinity based on the concentration of hydrogen ions (H$^+$). Blood plasma normally hovers around 7.4, which runs slightly on the alkaline side of the pH scale's neutral point of 7.0. The scale runs from 0–14. The blood pH, however, can vary based on the influx and efflux of hydrogen ions. Too many of these positive ions can shift the pH to a more acidic state (which is actually a lower number on the pH scale). Within the circulatory system, blood pH engages in normal swings. When the blood picks up carbon dioxide from the tissues for transport to the lungs, the carbon dioxide reacts with water to make carbonic acid (H$_2$CO$_3$), which quickly dissociates into carbonate ions (CO$_3^-$) and hydrogen ions (H$^+$). The extra hydrogen ions make the blood more acidic. That, in turn, can alter the hemo-globin's affinity for oxygen, making it more difficult for heme to bind to oxy-gen and thereby enhancing the release of oxygen from the hemoglobin. While this acidity-influenced discharge of oxygen works fine in blood that has already delivered the needed oxygen to the tissues, an acidic blood pH elsewhere, particularly in the lungs, could gravely impact oxygen delivery to tissues. This effect of pH on oxygen affinity is termed the **Bohr shift**. Some hemoglobin molecules remove hydrogen ions by binding to them to make acid hemoglobin, while the others continue to do their job unim-peded. In addition, plasma proteins similarly attach to hydrogen ions and act as buffers. In those cases where the acidity is too high to handle by any of these other means, hydrogen ions are relocated out of the blood and into the muscle tissue, which can withstand a temporarily higher pH. From this interim reservoir, the ions slowly flow back out of the muscles and into the blood, where they leave the body by way of acid urine or via exhalation from the lungs.

While the total amount of ions and proteins is important, the levels of in-dividual ions and proteins are also significant. For example, muscles and nerves respond to even slight changes in potassium and calcium ion concentrations. Likewise, cell membranes rely on the right combination of calcium, magnesium, potassium, and sodium in their immediate environ-ments.

Plasma proteins have other functions, too. One of their primary duties is transporting molecules from place to place. This job requires that the pro-teins have specific structures to which a hormone or other molecule can at-tach for its passage through the blood. For instance, hormones travel from the glands where they originate—the endocrine glands—to their final des-tinations throughout the body via the bloodstream. The glands secrete the hormones, then dump them directly into the blood, where plasma proteins latch on and provide swift passage to the target cells. This is an important service, because hormones are responsible for many normal functions, in-cluding the regulation of body metabolism, numerous aspects of reproduc-tion, responses to stress, and the body's growth and development. As an example, the pituitary gland, which is located at the base of the brain be-hind the eyes, produces a variety of hormones, each of which must travel the bloodstream to reach a wide range of tissues, including muscle, bone, skin, breast, ovary, and testis tissue, to take effect. (For more information on the operation of these organs and glands, see the Endocrine System volume of this series.)

Plasma proteins also transport metals, as well as many different types of medications. The plasma protein responsible for iron transport, for example, is known as transferrin, which is among the plasma proteins classified as globulins. In each case, the proteins simply act as taxi cabs. They don't interact directly with the hormones, drugs, metals, or other transported molecules besides handling the pickup and delivery.

As mentioned, plasma proteins include albumins, globulins, and fibrinogen. Fibrinogen makes up only 0.2 percent of plasma proteins, with the remaining 99.8 percent split as 55 percent albumins and 44.8 percent globulins. As previously described, the albumins are involved in maintaining blood volume and water concentration, and some of the globulins serve as transportation vehicles for a variety of molecules. Other important activities for globulins include blood clotting and immune responses. The immune response of plasma proteins will be discussed here, and blood clotting in the next section. (More details are available in the Lymphatic System volume of this series.)

Globulins come in three types: alpha, beta, and gamma globulins. The transferrin used in iron transport is a beta globulin. A variety of other beta globulins are collectively termed **complement**, and assist in the immune system by binding to passing potential antigens. Beta globulins identify invaders by telltale structures on the cell surface that are different than those of the body's own cells. Beta globulins are designed to grab onto these unusual structures, and they basically put a plug in the antigen's active site that renders it harmless. Similarly, beta globulins can ferret out cells that have antibodies already attached. In this case, another part of the body's immune response has already begun to mount a defense to the invading organism or foreign protein by creating the antibody. The beta globulins recognize the antibodies and attach to them. The captured cell then proceeds to the white blood cells for destruction.

Some plasma proteins, called **immunoglobulins (Ig)**, go a step further and act as antibodies themselves. The five main types of these immunoglobulin antibodies are:

- IgA, which is found in bodily secretions, like saliva, tears, milk, and mucosal secretions

- IgE, which causes the sniffing and sneezing associated with hay fever and asthma, and also defends against parasites

- IgG, which helps battle infections and also confers mother-to-fetus immunity

- IgM, common to almost every early immune response

- IgD, which has an unknown function

Proteins within each type of immunoglobulin are similar in basic structure, but have a variable region specific to a particular antigen. (Details are available in the Lymphatic System volume of this series.)

Another of the body's lines of defense depends in part on plasma proteins. Blood clotting requires fibrinogen, a soluble plasma protein; the beta globulins called **prothrombin** and **plasminogen**; platelets; and a slew of other molecules. The process is discussed in the next section.

PLATELETS AND BLOOD COAGULATION

Platelets, plasma proteins, vitamin K, and calcium all take their place in a quick-acting series of chemical reactions that result in blood coagulation, or clotting. Clotting begins almost immediately after the wound occurs as platelets congregate at the site of the injury. Platelets, also known as **thrombocytes**, aren't cells. Rather, they are sticky, disk-shaped fragments of large blood cells called megakaryocytes that reside solely in the bone marrow. These small (3.9–15.7 $\times$ 10^{-5} inches, 1–4 μm, in diameter) cell fragments exist throughout the circulatory system, with tens of millions in every droplet of blood. Their primary role is blood-clot formation. Because so many exist in the blood, a good supply of platelets usually isn't far from the wound site.

The first step in blood clotting is the release by the damaged tissue of a substance known as **thromboplastin**. As platelets arrive at the wound site, they disintegrate and release additional thromboplastin. Thromboplastin and calcium are both required to trigger the beta globulin called prothrombin to produce the enzyme thrombin. For the next step, the thrombin, platelets, and fibrinogen, a soluble plasma protein, work together to help make a tight web of insoluble fibrin threads that stick together and to the blood vessel wall. When blood cells encounter the web, they become trapped and form a blood clot. A scab is a dry, external clot. A bruise is a blood clot, too, but an internal one.

A number of other chemicals are also involved in the steps described above, and scientists are still trying to learn the details about them. New findings in 2002 indicate that a fatty molecule called lipid phosphatidylserine, or PS, is important to the conversion of prothrombin to thrombin. PS, which is located on the surface of platelets, stimulates the formation of another enzyme complex, called prothrombinase, that facilitates the prothrombin-to-thrombin transformation.

Although the coagulation process may seem complex, it happens very quickly. Small cuts are usually sealed within a couple of minutes, with an external scab hardening in place not long afterward. The yellowish fluid sometimes remaining at the injury site is called serum. Although the term serum is sometimes used interchangeably with plasma, serum actually refers to plasma that no longer contains fibrinogen or other clotting factors.

Normally, blood clots promote injury recovery by stopping blood loss, but that is not always the case. Clots that form within the blood vessels can be dangerous, because they can block blood flow and oxygen transport. A **stroke**, for example, is the result of a blood clot in the brain. (For more information on strokes, see Chapter 10.) Fortunately, platelets normally don't stick to the smooth walls of healthy, undamaged vessels. Other failsafes are heparin, which is found in basophils, and substances called antithrombins that turn off thrombin activity, effectively shutting down the coagulation machinery and preventing unnecessary blood clotting.

BLOOD TYPE AND RH FACTOR

The preceding introduction to blood cells may give the impression that blood in all individuals is alike. It isn't. The most obvious differences are blood type and Rh factor: Human blood types, or groups, are **type A**, **type B**, **type AB**, or **type O**, and Rh factors are defined as either positive or negative.

Blood type refers to the presence or absence of chemical molecules on red blood cells. These molecules can instigate antibody reactions and are therefore antigens. Red blood cells can have one, both, or neither of the two antigens named "A" and "B." Blood with only A antigens or only B antigens is called type A or type B, respectively. Blood with both A and B antigens is type AB, and blood with neither is type O. People with type A blood are also born with beta (or anti-B) antibodies, which are designed to detect and eliminate B antigens. Likewise, people with type B blood have alpha (or anti-A) antibodies that assail A antigens. Type AB blood has both antigens but neither antibody, and type O blood has neither antigen but both antibodies.

This confusion of letters means that type A blood donors can safely give their blood to any person who does not have antibodies to the antigens in their blood, namely A. As noted in Table 2.2, the type A donor's blood is compatible with the blood of recipients with type A or type AB. On the other hand, a person who has type A blood can receive blood donations from any person whose blood doesn't trigger a response from their own antibody contingent, beta. The table shows that type A persons can receive blood donations of type A and type O, because neither adversely reacts with the beta antibody in type A blood.

Reactions between mismatched blood can be severe. If type A blood from one person is given to another person with type B blood, the blood will clump due to a process called **agglutination**, as the alpha antibodies battle the B antigen. After clumping, the red blood cells will rupture in a process called **hemolysis**, which can lead to serious consequences, such as kidney dysfunction, chills, fever, and even death. For this reason, medical professionals compare blood type and Rh factor from a patient and a donor before proceeding with a transfusion.

Type AB-positive blood is frequently called the "universal recipient." Type AB blood has neither alpha nor beta antibodies, which means that any blood can be introduced without the chance of an antibody attack by the

TABLE 2.2. Blood Types

Blood Type	Antigens Present	Antibodies Present	Can Be Donated To	Can Accept Donations From
A	"A"	beta	type A, type AB	type A, type O
B	"B"	alpha	type B, type AB	type B, type O
AB	"A", "B"	none	type AB	type AB, type A, type B, type O
O	neither	alpha, beta	type O, type A, type B, type AB	type O

Blood type refers to the presence or absence of chemical molecules on red blood cells. These molecules, called antigens, can instigate antibody reactions. For this reason, medical professionals check blood compatibility before performing transfusions.

recipient's blood. The recipient doesn't have to worry about antibodies from the donor blood, because the amount of donated blood is small, becomes diluted in the recipient's blood, and presents no threat.

At the opposite end of the spectrum, type O blood is known as the "universal donor." It has neither the A nor B antigens, and therefore can be administered with little fear of agglutination. Nonetheless, medical professionals still take precautions to ensure blood-transfusion compatibility by mixing donor and recipient blood and watching it closely for adverse reactions. The reason for the wariness is that A and B aren't the only antigens. Sometimes, blood contains less common antigens that can bring about agglutination and cause problems for the patient.

One notable antigen is **Rh factor**, a protein so named because it was first discovered in Rhesus monkeys. People who have the protein, which is also known as **D antigen**, on the surface of their red blood cells have blood described as Rh positive, so a person with A antigens and Rh factor has type A-positive blood. A person with no A or B antigens, and no Rh protein has type O-negative blood. Unlike the A, B, and O types that come with associated antibodies, people without Rh factor are not born with the Rh antibodies. They can, however, produce Rh antibodies if one of two things happens. First, if an Rh-negative person receives a transfusion of Rh-positive blood, the Rh factor will stimulate the body to produce Rh antibodies. While the initial transfusion of Rh-positive blood is usually safe for the patient because antibody production is just starting up, any subsequent Rh-positive transfusion will be met with an army of antibodies and cause the clumping depicted previously.

The second circumstance that would result in Rh antibody formation involves pregnant women and their fetuses, and occurs when an expectant mother has Rh-negative blood, but her fetus carries Rh-positive blood. This occurs because Rh factor—as well as type A, B, AB, and O blood—is an inherited trait. It is possible, then, for a mother to have Rh-negative blood, while her fetus has inherited Rh-positive blood from the father. This usually presents no problems for the first child, provided the mother's blood has not already been sensitized to Rh-positive blood from a prior transfusion. The reason is that the circulatory systems of the fetus and mother usually are separated by a thin membrane. As the pregnancy comes to term, however, the membrane sometimes ruptures, allowing a small amount of fetal blood to enter the mother's bloodstream. This activates the maternal production of antibodies, which remain in wait for the next time her blood encounters Rh-positive blood. A second pregnancy with an Rh-positive child could threaten the fetus if some of the mother's blood crosses the membrane barrier and the antibodies attack the fetal red blood cells. More about this condition, known as **erythroblastosis fetalis**, is presented in Chapter 10.

BLOOD CELL RENEWAL AND PRODUCTION

Blood is a renewable resource. Red and white blood cells, platelets, and other components are constantly turning over. For example, the average red blood cell lives about four to six months, and more than 2 million red blood cells die every second. New red blood cells are being generated just as fast, however, so the concentration of erythrocytes in the blood remains quite constant. The same death and replenishment cycle occurs with other blood cells. As stated earlier, however, white blood cell numbers can quickly rise, if necessary, to fight infections.

The **bone marrow**—spongy, connective tissue inside many bones—is the place where adult blood cells and platelets typically originate. If a person's marrow is damaged, cells in the liver and spleen can make red blood cells. Cells in the body reproduce in two basic ways. Some types of cells originate through simple duplication where the parent cell divides to produce two new "daughter" cells. This process is called **mitosis**. Other cells, like the blood cells, arise from stem cells. **Stem cells** are undifferentiated, which means they have the genetic potential to become one of any number of distinct cells. In other words, stem cells are the precursors to the final cells.

Some are unipotent, or "hard-wired," to eventually develop into only one particular type of cell. Pluripotent stem cells, alternatively, can ultimately mature into more than one specific end-product cell. The process amplifies because stem cells can divide again and again into daughter cells that will either remain stem cells themselves or differentiate into specific cells.

Various factors, especially specific hormones, incite daughter cells to begin the path toward an end-product cell. This process is irreversible. For example, a stem cell cannot mature into a red blood cell, then revert into a stem cell and switch into a white blood cell. The exact process and all of the minute details are still under study, but the basic pathways for **pluripotent hemopoietic (blood-forming) stem cells** are described in Figure 2.4. In

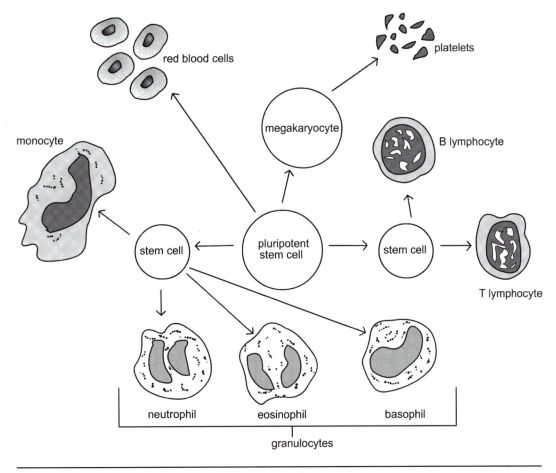

Figure 2.4. A simplified view of blood cell development.
Several other unpictured developmental stages are involved in the development of blood cells. In the case of red blood cells, for instance, the stem cells first differentiate into large, nucleated cells. As they mature, they gain hemoglobin, lose their nuclei, become smaller, and take on their characteristic biconcave shape.

a normal human being, these stem cells are an unending source of red and white blood cells, and platelets in the bloodstream.

In the first major step of red blood cell development, stem cells differentiate into **erythroblasts**, which are about twice the size of mature red blood cells and have nuclei. As the forming erythrocytes mature, they gain hemoglobin but lose their nuclei. They also become smaller and take on their typical biconcave shape as a round disk with a central depression on each side. The last stage before becoming a fully formed red blood cell is called a **reticulocyte**. About 1–2 percent of red blood cells in the bloodstream are reticulocytes, and the remainder are fully mature erythrocytes. The average lifespan of a red blood cell is 120 days.

Because platelets are actually fragments of megakaryocytes, their evolution begins when stem cells embark upon the pathway to megakaryocyte formation. Megakaryocytes are large, ranging from 1.6 to 3.9×10^{-3} inches (40 to 100 μm) in diameter, and have odd-shaped, many-lobed nuclei. The megakaryocytes are then broken into much smaller end-product platelets that average just 0.12×10^{-3} inches (3 μm) in diameter. The fragmentation occurs in the marrow, lungs, and sometimes the spleen. Platelets last between four and ten days.

Among the white blood cells, the neutrophils, eosinophils, basophils, and monocytes complete their development in the marrow. Lymphocytes, on the other hand, mature in the thymus and liver as well as in the marrow. For example, B lymphocytes typically mature in the bone marrow, while T cells finish their development in the thymus. Some then enter the bloodstream, while others remain in the **lymphatic system**, which is a series of tubules that shunts excess tissue fluid into the veins through two ducts. (More information is available in Chapter 7. For additional details, see the Lymphatic System volume of this series.)

As shown in Figure 2.4, all of the granulocytes (neutrophils, eosinophils, and basophils) share a few developmental steps. From the stem cells, they differentiate into myeloblasts and then progranulocytes before splitting into separate pathways. The lymphocytes and monocytes each head in their own direction in the earliest stages of development. Considerable variability exists among white blood cell lifespan as well. The "typical" leukocyte lives only about two weeks, but some can exist for 100–200 days or more.

As this chapter has demonstrated, blood is crowded with cells and other substances that are involved in many functions critical to life. The next chapter will consider the vessels that carry the blood throughout the body.

Artificial Blood

As blood banks have cried out for donors, and scares about contaminated blood have surfaced in connection with AIDS and other diseases, scientists have been developing artificial blood that is both safe and plentiful.

Studies into artificial blood had their start in the 1960s when researchers began animal experiments using liquids containing oxygen-absorbing fluorocarbons. In 1979, scientist Ryochi Naito of Japan tried the concoction on himself and survived a half-pint (200 ml) transfusion of the liquid.

Since then, researchers have refined artificial blood. Many of these blood alternatives are solutions of hemoglobin, the molecule that delivers oxygen to body tissues and removes carbon dioxide. In some cases, the hemoglobin is obtained from cow's blood or from stored, donated human blood that is too old to be transfused. This artificial blood contains free hemoglobin—in other words, it isn't packaged into red blood cells as it normally is in human blood. While these artificial hemoglobin solutions performed their primary functions, they appeared related to hypertension and other problems.

By the late 1990s, researchers were developing artificial blood that solved some of these problems. For example, a team at Baxter Hemoglobin Therapeutics Inc. in the United States found that the cell-free hemoglobin molecules seemed to cause the blood vessels to contract, which leads to hypertension. They hypothesized that the contraction resulted from nitric oxide binding to the hemoglobin, and devised a hemoglobin molecule to discourage such binding. Tests in rats concluded that the new hemoglobin reduced high blood pressure.

Several companies are now testing and beginning to market artificial blood made with free hemoglobin. Some, like Sangart Inc. in San Diego, are refining artificial blood products to make them last longer in the bloodstream. Their work has the potential to boost the lifespan of the products, which is typically only a few days. Other research groups are employing genetic engineering or developing completely synthetic artificial blood. For example, scientists at Carnegie Mellon are putting the genetic instructions for hemoglobin into the common gut bacterium called *Escherichia coli* (*E. coli*), which allows the bacterium to make hemoglobin. Researchers at Alliance Pharmaceuticals in San Diego, on the other hand, are using long chains of carbon and fluorine atoms that are able to carry oxygen and carbon dioxide.

While this research is proceeding, others are considering the possibility of converting embryonic stem cells into blood cells. Stem cells are the undifferentiated cells from which all other types of cells arise. In the September 11, 2001 issue of *Proceedings of the National Academy of Sciences*, a research group from the University of Wisconsin–Madison indicated that it had already been successful in converting the stem cells into erythrocytes, leukocytes, and platelets. Studies are now under way to determine if these blood cells are robust enough to survive transfusion.

3

Blood Vessels: The Transportation System Within

As previously described, the blood isn't just red liquid. It is filled with millions of cells, each with a specific job to do. Likewise, blood vessels are much more than a simple series of pipes to contain and route blood. Blood vessels are living, dynamic tissues with complexities all their own.

The circulatory system in a single adult human being comprises some 60,000 miles of blood vessels. Most people can see at least a few of them just under the skin of the wrist. The vast majority are much smaller than those visible vessels, and have diameters of less than three-thousandths of an inch (about 10 microns). Blood vessels are associated with one of three major groups: the arterial system, the venous system, or the capillary system. Whenever blood is moving away from the heart, either to the body tissues (systemic circulation) or to the lungs (pulmonary circulation), the arterial system is involved. The capillaries take over when the blood reaches its destination, and serve as the exchange vessels between the blood and the lungs, or between the blood and the body tissues. When the exchange is complete, the blood moves from the capillaries into the vessels of the venous system, which directs the blood back to the heart to begin another route either to and from the lungs, or to and from the body tissues.

All three types of vessels, then, participate in the circuitous path of the blood from the heart to the lungs and back to the heart, and from the heart to the other body tissues and back. The heart-to-lungs-to-heart path is called the **pulmonary circulation** and serves to allow the blood to pick up oxygen. The heart-to–body tissues–to-heart circuit is called the **systemic circulation** and allows tissues to take up oxygen and other materials transported in the blood. (Occasionally, some texts separate the circulation within the heart from the systemic circulation and call it the **coronary circulation**.) Overall, the blood travels more or less in two loops, one from the heart to the lungs and back, and a second from the heart to the body tissues and back to the heart.

ARTERIAL SYSTEM

The main function of the **arterial system** is to carry blood away from the heart and either to the lungs to pick up oxygen, or to other body tissues to drop off nutrients, oxygen, hormones, or other needed substances. Like a river system that has main branches from which diverge many smaller side creeks, the arterial system has main branches called arteries and many diverging, smaller vessels called arterioles (see color insert). The major arteries provide quick, direct routes to major body areas, and smaller arteries divert blood to more specific sites. Separating from the arteries, the arterioles bring blood to specific target tissues. The arterial system works much as a road system does: travelers use superhighways to get quickly to a general region, then take smaller freeways and finally side roads to reach a specific destination. In the case of the arterial system, the blood moves along main arteries, then into smaller arteries and even smaller arterioles. The specific destination of the blood cells is a set of capillaries in the lungs or in some other body tissue.

Vessel Composition

Arteries and arterioles are more complicated than they might seem at first, and that complexity begins with the structure of the vessels themselves. Arteries and arterioles are made of three concentric layers: the tunica adventitia, the tunica media, and the tunica intima (see Figure 3.1).

The **tunica adventitia** coats the outside of an arterial vessel and is made of fibrous connective tissue that loosely holds the vessel in place. In the larger arteries, the adventitia also holds a number of small blood vessels of its own. These small vessels, called **vaso vasorum**, provide nourishment to the thick vessel walls.

The thickest of the three layers is the **tunica media**, which is the muscular and elastic middle layer of vessel walls. This layer allows the large arteries to expand and contract in tune with the waves of blood accompanying each beat of the heart. Widening slows the blood flow and narrowing quickens it, so the combined action helps to moderate the blood's speed. **Elastin**, a protein that has six times the spring of rubber, provides the vessels' elasticity and allows the vessels to stretch wide as a pulse of blood arrives. A much stiffer protein, called **collagen**, prevents the vessels from expanding

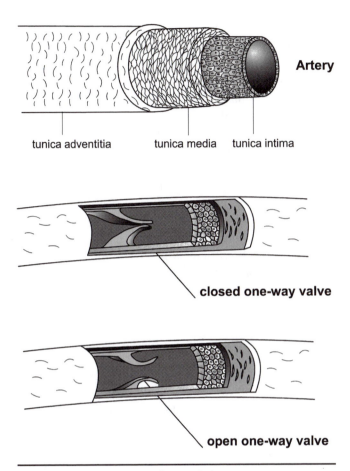

tunica adventitia   tunica media   tunica intima

Artery

closed one-way valve

open one-way valve

Figure 3.1. Arteries and arterioles.
Arteries and arterioles have three layers. The outer layer, called the tunica adventitia, is made of fibrous connective tissue and loosely holds the vessel in place. The middle, or tunica media, is a thick, muscular, elastic layer. The inner layer, or tunica intima, is a one-cell-thick layer of endothelial cells. Many of the larger blood vessels contain valves that open when blood is pulsing forward and close against any back flow. This ensures a unidirectional blood flow.

too much. Stretching is essential for arteries lying close to the heart, where the force of the heart's pumping on blood flow is most strongly felt. Here, arteries distend to take the brunt of each blood rush, then recoil to create a more even blood flow to subsequent areas of the circulatory system. As might be imagined, these arteries contain a higher percentage of elastin than other vessels further down the line in the systemic circulation.

For the contraction of arteries, the spindle-shaped, smooth muscle cells of the tunica media take over. Overall, the body has three types of muscle:

smooth, striated, and cardiac. **Striated muscles** are those that a person can consciously control. Leg muscles are an example. A jogger or walker can control the muscles to speed the pace or to slow down. **Smooth muscle**, on the other hand, is controlled mainly by the **autonomic nervous system**, which means that these muscles work involuntarily or outside of a person's will. Unlike striated muscle that tires rather quickly (as any jogger will attest), smooth muscle can continue working for long periods of time. **Cardiac muscle** has characteristics of both smooth and striated muscle, and will be discussed in Chapter 5. The tunica media, or middle layer of the vessels, has smooth muscle, which wraps around the vessel rather than running its length and is responsible for contracting arteries. When the smooth muscle tightens and decreases a vessel's diameter, blood pressure rises because the blood is forced through a narrower opening. The effect is similar to that achieved by holding the thumb partially over the water flow at the end of a garden hose. The decreased diameter of the hose at the point of constriction (the thumb) increases the water pressure. As stated earlier, the arteries closest to the heart have the most elastin to slow the strong pulses of blood leaving the heart. They also have the least smooth muscle. In contrast, vessels farther away from the heart have a higher proportion of smooth muscle to control vessel diameter and help keep the blood moving.

The innermost layer of the blood vessel is the **tunica intima**, which is also known as the **endothelium**. This layer has direct contact with the blood that runs through the artery or arteriole. Although it is only one cell thick, the tunica intima's flat and smooth cells are important in imposing a barrier of sorts and preventing the passage of plasma proteins out of the blood. In arteries that connect with the heart, the tunica intima has a thin, fibrous layer that blends seamlessly with the heart's inner lining, known as the **endocardium**.

Systemic Circulation

On the arterial system's systemic-circulation side (see color insert), which transports blood from the heart to the body, the first artery to leave the heart is also the largest one. It is called the **aorta** and has an internal diameter of about 1 inch (about 25 mm) in the average adult. This internal diameter is called the **lumen** and represents the open space in a vessel through which the blood flows. The walls of the aorta are about 0.079 inches (2 mm) thick. In comparison, other arteries average about 0.157 inches (4 mm) in internal diameter with a wall thickness of about 0.039 inches (1 mm). Arterioles are commonly classified as arterial vessels with overall diameters of less than 3.9×10^{-3} inches (100 μm). An average arteriole has a lumen of about 1.2×10^{-3} inches (30 μm) and a wall thickness of 7.9×10^{-4} inches (20 μm). Capillaries range from about 1.6×10^{-4}–3.9×10^{-4} inches (4–10 μm) in internal diameter with a wall thickness of just 2.0×10^{-5}–3.9×10^{-5} inches (0.5–1 μm).

The blood entering the aorta from the heart has just been to the lungs, so it is fully oxygenated and bright red. As just described, the aorta is highly elastic and can distend greatly to accept the powerful rush of blood pumped from the heart. As the aorta returns to its original size, the blood moves out in a more even flow. Although the blood eventually flows very smoothly, the pulse can be felt by pressing on some areas of the body where main arteries run close to the skin surface. For example, nurses typically determine a patient's heart rate by feeling the wrist, or the radial pulse site, and counting the number of pulses over a set period of time. A normal resting adult's heart rate is about 70 beats per minute. The rate is typically lower in adult athletes and higher in children. Other major arterial pulse sites include:

- Temporal, in front and slightly above the ear
- Facial, along the lower jaw
- Carotid, beside the windpipe in the neck
- Brachial, on the inside of the elbow
- Femoral, on the upper thigh beside the groin
- Popliteal, on the back of the knee
- Posterial tibial, on the inner ankle
- Dorsal pedal, just above the toes on the upper foot

The aorta is a large, arched artery that originates at the heart, where it connects to the lower left heart chamber, called the **ventricle**. From there, it continues down the trunk of the body. All of the major arteries in the human body branch off of the aorta. Using the analogy of a road system, the aorta would be the main superhighway through which all outgoing (systemic) traffic has to pass. The highways that divert from it would represent the major arteries. These major arteries supply blood to all of the main body regions, including the limbs, head, and body organs. The arterioles are the side roads that bring the blood to specific locations throughout the body.

Two major arteries parting from the **aortic arch** are the right and left **coronary arteries**, which provide blood to an important body organ, the heart. These two arteries further subdivide. The left coronary artery splits into a circumflex branch that runs behind the heart and an anterior descending branch that curves forward over the heart. These two branches mainly supply the left ventricle and left **atrium**, which is the heart's left, upper chamber. The right coronary artery and its posterior descending branch deliver blood mainly to the right side of the heart, which has its own ventricle and atrium. In addition, both the left and right coronary arteries supply blood to the **septum**. The septum is the interior wall between the right and left ventricles.

All of the blood pumped from the heart into the aorta in one contraction is called the **total cardiac output** or **stroke volume**. Because the heart is such a hard-working organ, it receives a rather large portion, about 5 percent, of the total cardiac output even when a person is resting. As will be discussed later, that proportion can change if a person is active or under some form of psychological or physical stress.

Blood travels to the head via two major vessels called the left and right **common carotid arteries**. The left carotid splits directly from the aortic arch between the bases of the two coronary arteries. The right carotid indirectly branches from the aorta via a short vessel, called the **brachiocephalic** (or **innominate**) **artery**. The brachiocephalic artery also feeds the right subclavian artery, which is discussed below. Each of the two carotids splits into internal and external carotid arteries, which are the principal arterial vessels of the head and neck.

Other arteries that branch from the aorta soon after it leaves the heart include the right and left **subclavian arteries**. Like the right carotid artery, the right subclavian artery divides off of the brachiocephalic artery. Each subclavian artery supplies an arm. They earned the name subclavian because their paths to the arms run beneath the collarbone, or clavicle. Blood flow continues down the length of the arm through the **brachial artery** and into the **ulnar arteries** and **radial arteries** of the forearm.

As the aorta travels down the spine, it is called the thoracic aorta in the chest, or thorax, and then becomes the abdominal aorta. Along the way, it supplies blood to various internal organs, such as the kidneys, spleen, and intestines. Each of the two kidneys, for example, gets its arterial blood supply from a **renal artery** that separates from the aorta. The renal artery splits into smaller and smaller vessels, eventually leading to a round cluster of capillaries, which is called the **glomerulus**. Fully 25 percent of the total cardiac output goes through the renal arteries to the pair of kidneys. These two organs not only excrete waste products through the urine but also have important roles in regulating the amount of water in the blood, as well as its pH level (a measure of acidity/alkalinity) and electrolyte content.

Some of the other major organs and systems receiving blood from the aorta—in this case, the abdominal artery—are the spleen, which is fed by the **splenic artery**; the liver, which gets its blood via the **hepatic artery**; and the digestive system, which is supplied by a number of arteries, including the **gastric**, **gastroepiploic**, superior and inferior **mesenteric**, sigmoidal, and **colic arteries**. Additional information on circulation in major organs is addressed in Chapter 7.

Many of the arteries' names come from the medical names of their destination. The renal artery is named for the renal, or kidney, system; the ulnar and radial arteries ship blood to the area surrounding the ulna and radius, which are bones in the forearm; and the superior mesenteric artery supplies the region around the intestinal membrane, which is known as the mesentery.

The abdominal artery ends when it bifurcates into the left and right common **iliac arteries**, each of which soon divides again into internal and external iliac arteries (sometimes called hypogastric arteries). From each of the internal iliac arteries come various arteries that supply blood to the pelvic area, including the reproductive organs. The external iliac artery becomes the **femoral artery** when it enters the thigh.

Continuing with the leg, the femoral artery of the thigh becomes the **popliteal artery** at the knee (popliteal is a medical term for the back of the knee), then divides into the posterior and anterior **tibial arteries** (the tibia is a bone in the lower leg).

These are just some of the major named arteries in the human body. A closer examination will reveal a host of smaller and smaller arteries that branch from these major arteries, as well as the array of arterioles that feed the capillaries. The arterioles have an important role in regulating the amount of blood entering the capillaries. Using their layer of smooth muscle, the arterioles constrict or dilate to increase or decrease the blood flow. For this reason, arterioles are sometimes called **resistance vessels**. Once the blood moves into the vast web of capillaries, the oxygen and nourishment they are carrying are relinquished to the target tissues.

As described previously, the arterioles-to-capillaries-to-venules route is the most common pathway for blood. However, the arterioles in a few tissues never connect with capillaries, instead attaching to venules by way of wide vessels called **arteriovenous anastomoses**. These muscular vessels, which range from 7.9×10^{-4}–5.3×10^{-3} inches (20–135 μm) in diameter, are common in the skin and in the nasal mucosa (in the nostrils), and regulate body temperature. Heat from the body's core is transported via the blood to these areas for release to the outside.

Before exploring the capillary system, this chapter will examine the pulmonary circulation, which takes blood from the heart to the lungs for oxygenation and back to the heart.

Pulmonary Circulation

Pulmonary circulation is less complex than systemic circulation because the only target organ for the arterial system is the lungs (see color insert). There, the blood picks up the oxygen that it will eventually deliver to the body through the systemic circulation, which was just discussed.

Unlike the systemic arterial circulation, which heralds from the heart's left ventricle at the aorta, the pulmonary arterial circulation begins with the right side of the heart and the **pulmonary artery**. The pulmonary artery connects to the right ventricle, which is the smaller of the heart's two ventricles. As the right ventricle contracts, it pushes blood into the pulmonary artery. The artery soon splits, and its two branches lead to either the right or left lung.

Just as blood in the systemic circulation moves from major to smaller arteries, and then to arterioles and capillaries, the blood in the pulmonary system diverts into smaller and smaller vessels, ultimately ending at the capillaries. Once the blood enters these tiny vessels, it picks up the molecules of oxygen that are so important for cellular function. That process will be described in the section on the capillary system.

VENOUS SYSTEM

On many levels, the venous system is the opposite of the arterial system (see color insert). On the systemic side, the arterial system delivers blood away from the heart and to the tissues, and the venous system goes the other way, bringing the blood back to the heart. On the pulmonary side, the arterial system sends blood from the heart to the lungs for oxygenation, and the venous system sends the now oxygen-rich blood back to the heart. Revisiting the road analogy, a traveler might leave a major city (the heart) via a superhighway (the aorta), then take a smaller highway (the arteries) and finally side roads (the arterioles) to get to a small town (the tissues) or other destination. To return to the city, the traveler would go in the opposite direction, beginning by taking the side roads, which are analogous to the venules; then the highways, or the veins; and finally the superhighway that leads into the city, or the heart. The venous system has two superhighways, which are two large veins called the superior **vena cava** and the inferior vena cava, which will be described later. (The plural of vena cava is venae cavae.)

Besides its role in returning blood to the heart, the venous system is a blood reservoir. Typically, about two-thirds of the body's circulating blood supply is in the venous system. The veins and venules temporarily store blood that can be immediately transported to the other areas of the body as necessary. Exercise, for example, initiates a series of responses within the body that affect the circulatory system. One occurs when **vasoconstrictor nerves** send messages primarily to the veins that tell them to constrict. As they do, blood moves from the venous system to the heart, which in turn sends added oxygenated blood to the arterial system and facilitates the increasing need for oxygen in the working muscles. Similarly, when a person loses a large volume of blood through a serious wound, nerves signal a reorganization of the blood from tissues that are less important to immediate survival to those tissues that are vital in maintaining life. The venous vessels constrict in some areas, such as the skin or the digestive tract. This forces blood to vessels in vital organs, like the heart. The action not only preserves oxygenation to critical tissues but also helps maintain the body's overall blood flow and ensure that the blood reaches its destination without delay.

Another important difference between the venous and arterial systems is the movement of the blood. Blood efficiently flows through veins and venules even though the driving **blood pressure** is lower than it is in the arterial system. Blood pressure is measured in a unit abbreviated as mmHg. It refers to millimeters (mm) of mercury (Hg), the standard method of measuring pressures. Scientists measure pressure by watching its effect on mercury—the silver liquid in old-fashioned thermometers. If mercury is placed in a tube, an increase in pressure will cause the mercury to expand and rise. A pressure decrease results in a drop of the mercury level. A reading of 120 mmHg means that the pressure in the aorta is enough to raise a column of mercury by 120 mm above the zero point, which is normal atmospheric pressure.

In the venous system, just 10–15 mmHg pressure is enough to force blood all the way from the venules back to the heart. This is about half of the average pressure of 30 mmHg seen in even in the small arteries. The venous system can function on the lower pressure, which nears 0 mmHg (atmospheric pressure) by the time it gets to the heart, in part because the branching of the venous system is going in a direction opposite to that of the arterial system. Instead of an aorta that branches into smaller arteries and then miniscule arterioles—and slows as it goes, the venous "river" does the opposite. Blood from perhaps dozens of tiny venules merges into a slightly larger venule. When it does, the blood rate increases because the overall lumen (interior diameter) of the larger venule is still less than the combined lumens of all the smaller venules feeding into it. In other words, blood from a bigger space is squeezing into a tighter space. The same thing happens when venules merge into one small vein, or when numerous veins merge into the vena cava. The flow accelerates along the way much as a river's current hastens as new creeks join it.

The elasticity of the large and small venous vessels is also opposite to that of the arterial vessels. Whereas the largest arteries are the most elastic to tame the pulses of blood from the heart, the largest veins are the most rigid to help maintain or boost the blood flow. Conversely, the most distensible vessels on the venous side are the venules, which readily expand to serve as blood reservoirs.

The venous and arterial systems have other similarities and dissimilarities, too, and these will be discussed in the next sections.

Vessel Composition

Veins and venules have a three-layered structure much like that of arteries and arterioles (see Figure 3.2). Each of these vessels has the same three layers: the tunica adventitia on the outside, the tunica media in the middle, and the tunica intima lining the inside. Just as in the arterial vessels, the muscular and elastic tunica media is the thickest layer in veins and venules, and includes both elastin and muscle tissue to allow the vessel openings to expand and constrict. The elastin protein gives the vessels their stretch and slows the blood moving through them, while the smooth muscle cells contract and narrow the vessels' openings to urge the blood's pace. Collagen in this middle layer helps to rein in the vessels' elasticity and yields structural strength.

The tunica adventitia, or outer layer, of veins and venules is a covering of connective tissue. Its job is to hold the vessel in place. The innermost of the three layers, or the tunica intima, is a one-cell-thick sheet of endothelial tissue.

So far, this description of the composition of veins and venules is the same as that for arteries and arterioles, but differences do exist. In the arterial system, the delineations between the three layers are much more distinct than they are in the venous system. The transition in venules and veins is rather gradual. In the arteries and arterioles, the smooth muscle cells of the tunica media wrap around the vessels in a very regularly arranged fashion. The veins and venules have a thinner middle layer with fewer muscle cells that are commonly in a less-ordered arrangement. This would result in a severe lessening of their mechanical strength were it not for their high

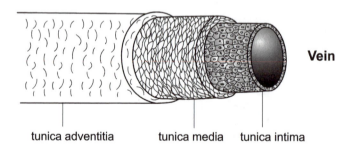

tunica adventitia   tunica media   tunica intima

Figure 3.2. Veins.
Veins and venules have a three-layered structure: the tunica adventitia on the outside, the tunica media in the middle, and the tunica intima lining the inside. The main difference between these vessels and those of the arterial system is in the middle layer. The veins and venules have a thinner tunica media with a higher collagen content and with fewer muscle cells that are commonly in a less-ordered arrangement.

collagen content. The additional collagen typically seen in veins and venules also limits the vessels' elasticity somewhat. As seen in the arterial system section, arterial vessels, especially the larger arteries, must distend to even out the strong blood pulses from the heartbeat and regulate the blood flow. There is no similar demand on veins and venules, but they do need to stretch out enough to accommodate the temporarily stored blood they are holding. In effect, veins and venules are striking a balance between stretching to hold up to 70 percent of the body's blood and remaining inelastic enough so that they can maintain a sufficient flow to transport the blood back to the heart.

Venous blood faces other challenges on its way back to the heart. The force of gravity discourages blood flow. In a standing individual, the blood in the feet has to overcome the gravitational pull to flow up the leg and back to the heart, and it does it without the heart's pumping action to help it along. While fighting gravity, the blood also must flow into increasingly bigger vessels, which would seemingly have the effect of slowing blood flow. In large part, blood can overcome these obstacles for the same reason that venules develop a flow with the influx of capillaries: Blood is still moving from a larger space to a smaller space. Because so many smaller vessels feed a larger vessel, the combined cross-sectional area of the smaller vessels is greater than that of the larger vessel, and blood flow actually speeds up as it approaches the heart. In addition, venous vessels become stiffer as they get closer to the heart, because their collagen content is greater. This ensures that the vessels won't distend and therefore won't create a larger space for blood or slow the blood rate. (Exercise and other stressors can accelerate the venous blood flow, and this will be discussed in Chapter 5.)

Systemic Circulation

The systemic side of the venous system begins at the capillaries and ends at the heart (see color insert). The capillaries are the site of blood-to-tissue transfer of nutrients and other materials, including oxygen. Once that oxygen and nutrient transfer has occurred at the tissues, the venous system takes over to collect the blood from the capillaries and convey it back to the heart. Oxygenated blood in the arterial system is bright red, because oxygen causes a change in the three-dimensional configuration of the large compound called hemoglobin that is found in red blood cells. Without the oxygen, the blood appears dark maroon. For this reason, blood leaving the capillaries—after the oxygen drop-off—is more blue than red.

The venules are the first vessels in the return of blood from the capillaries to the heart. A typical venule has a lumen of about 7.9×10^{-4} inches (20 μm) with a vessel wall that is about 7.9×10^{-5} inches (2 μm) thick. This compares to the average arteriole, which has a like-sized lumen but a wall thickness of 5.9×10^{-4} inches (15 μm). Many capillaries, which are about a quarter of the size of a venule, may empty into a venule, and, as already described, this onslaught helps to generate an increased flow. From there, smaller venules merge into larger and larger venules, and eventually into small veins. Although the size varies considerably, a typical vein has a lumen of around 0.197 inches (5 mm, or 5,000 μm). Its wall thickness averages about 0.02 inches (0.5 mm, or 500 μm). When looking at the circulatory system as a whole, the venous vessels often have arterial counterparts, with blood flowing out to the tissue in an artery and back to the heart in a nearby, sometimes adjacent, vein. For an overview of the major vessels in both the arterial and venous systems, see the map of the major arteries and veins in color insert.

This chapter will now provide a closer view of some of the major veins in different body areas, beginning with the leg. Blood from the foot may ascend into the anterior **tibial vein**, which is named for the tibia (one of the lower leg bones). Veins in the ankle and numerous capillaries in the leg empty first into **peroneal veins**, which are also known as **fibular veins** because of their location in the region of the fibula (the other lower leg bone), and then into the posterior tibial vein, which eventually unites with the anterior tibial vein. Both the anterior and posterior tibial veins, which also accept blood from numerous other capillaries in the lower leg, flow into the **popliteal vein** at the back of the knee. The popliteal vein, in turn, empties into the **femoral vein**, a large vessel in the thigh (alongside the large upper-leg bone, or femur). Blood from the foot can also take a more direct route to the femoral vein by way of the **great saphenous vein**, the longest vein in the human body.

Regardless of how it reaches the femoral vein, all of this blood flows into the external **iliac vein**. Now in the abdomen, blood from the external iliac vein joins with the internal iliac vein, which carries blood from the pelvis to form the common iliac vein. This vein joins the large inferior vena cava, one of the two "superhighways" in our road system bringing blood back to the heart.

Many other major veins in the abdominal cavity empty directly or indirectly into the inferior vena cava. Within the blood supply for the reproductive system, for example, the female body has a pair of **ovarian veins**, and the male body has a pair of **spermatic veins**. Both the right ovarian vein and the right spermatic vein connect directly to the inferior vena cava, but the left ovarian and spermatic veins first merge with the left **renal** (kidney) **vein**, which then unites with the inferior vena cava. The female's **uterine veins** take a more convoluted route to the inferior vena cava, first merging with the internal iliac vein that unites with the common iliac vein, and finally joining the inferior vena cava.

Other major body areas that use the inferior vena cava include the kidneys, liver, spleen, digestive system, and pancreas. The renal veins of the kidneys and the **hepatic veins** of the liver empty directly into the inferior vena cava. The liver performs a vital function because it absorbs products that the blood has gained from the digestive system. For example, the liver absorbs glucose (a sugar that results mainly from starch digestion) and uses much of it to make glycogen, which is basically a storable form of glucose. When the body needs extra energy, the glycogen transforms back into glucose.

Blood from the spleen takes a less-direct route to the inferior vena cava. It drains from the spleen via the **splenic vein**, which joins the superior **mesenteric vein** to create the **portal vein**. The portal vein empties into the liver. Venous blood leaves the liver through the hepatic veins, as previously described, and discharges into the inferior vena cava.

The digestive system has many veins emptying the intestines, rectum, stomach, and other specific areas. These veins include the **rectal**, pudendal, **lumbar**, superior and inferior mesenteric, gastric, gastroepiploic, and **epigastric veins**. Each has its own path to the inferior vena cava. As an example, the rectal veins number three: the inferior, middle, and superior rectal veins. The inferior rectal vein joins the internal pudendal vein, which flows into the internal iliac vein, while the middle rectal vein connects directly to the internal iliac vein. The internal iliac vein then continues to the common iliac vein, which finally unites with the inferior vena cava. The superior rectal vein avoids the internal iliac vein altogether and instead drains into the inferior mesenteric vein, which flows into the splenic vein, then to the superior mesenteric vein and portal vein, through the liver, into the hepatic vein, and to the inferior vena cava. For a detailed description of the venous system in various major organs, see Chapter 7.

As the inferior vena cava returns blood from the lower body to the heart, another superhighway is doing the same for the upper body. This large vein is the superior vena cava, which gathers blood from the arm, chest, head, and neck regions.

The arm's arrangement is somewhat similar to that of the leg. The veins, of course, are named differently to reflect the specific body area in which they are found. Veins in the hand flow into the **radial vein**, the **ulnar vein**, or the **basilic vein**. The radial vein eventually merges into the ulnar vein, which then continues into the **brachial vein** in the upper arm. The basilic and brachial veins flow into the **axillary vein** that carries blood into the chest. Just as some blood from the foot can patch nearly directly into the femoral vein via the long saphenous vessel, some of the blood from the hand can drain through a long **cephalic vein** right to the axillary vein of the upper arm. The axillary vein flows into the **subclavian vein** and then the **brachiocephalic vein** (also called the **innominate vein**) of the upper chest.

The head and neck region have several major veins, but the most well-known are the **jugular veins**, which accept blood from the brain, face, and neck. These veins flow into either the subclavian or brachiocephalic veins. One of the three jugular veins, the internal jugular vein, is the largest venous vessel in the head and neck. This vessel actually merges with the subclavian vein to form the brachiocephalic vein of the upper chest.

Blood from tissues of the chest muscles, from the thyroid gland, and from the diaphragm also either directly or indirectly release into the brachiocephalic vein. The brachiocephalic vein drains into the superior vena cava.

The ultimate destination of both superhighways—the inferior vena cava and the superior vena cava—is the heart. Specifically, they deliver blood to the right atrium of the heart. The blood's movement through the heart will be discussed in Chapter 5.

Pulmonary Circulation

Like the arterial system, the pulmonary side of the venous system is much simpler than the systemic side (see color insert). Just two major veins are involved: the **bronchial vein** and the **pulmonary vein**. Most veins are single or paired, but the pulmonary veins are actually four vessels that flow directly from the lungs to the left atrium. These veins return newly oxygenated blood from the lungs to the heart. The bronchial veins, on the other hand, drain blood of the bronchi and a portion of the lungs, then travel through one or more smaller veins to reach the superior vena cava, which brings blood into the heart's right atrium.

CAPILLARY SYSTEM

Although the capillaries are the smallest vessels in the circulatory system, they represent the main exchange site between the blood and the tissues. They can be viewed as both the ultimate destination of the arterial system and the starting point of the venous system. From the heart, blood travels through the arteries to the arterioles, and then to the capillaries, where exchange occurs. Nutrients, oxygen, and other materials carried by the blood are traded for waste products from tissue cells. Blood continues down the capillaries, soon entering the venules and then the veins on its return trip to the heart.

Vessel Composition

Capillaries have a composition that is somewhat different from that of other circulatory vessels. As already described, arteries, arterioles, veins, and venules have a three-layered construction, including elastic tissue and smooth muscle. Capillaries, in contrast, are composed of just a single layer of endothelial cells, which gives them a wall thickness of just 2.0×10^{-5}–$3.9 \times$

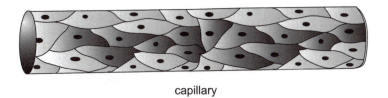

capillary

Figure 3.3. Capillaries.
Capillaries have an important function in exchanging gases and materials between blood and tissues, but their composition is quite simple. A capillary is basically a tube comprised of a single layer of epithelial cells.

10^{-5} inches (0.5–1 μm) (see Figure 3.3). Their internal diameter, or lumen, is about 1.6×10^{-4}–3.9×10^{-4} inches (4–10 μm). That's a tight fit for red blood cells, which range from 2.6×10^{-4}–3.5×10^{-4} inches (6.5–8.8 μm) in diameter. To squeeze their way through these tiny vessels, red blood cells travel in single file (a so-called bolus pattern), and when that's not enough, they bend, twist, partially fold, and otherwise deform. At the same time, the capillaries distend to allow the blood cells to pass through them. It is a tortuous pathway, but it is a short one: Capillaries are typically just 0.02–0.039 inches (about 0.5–1 mm) long.

Capillaries come in three types:

- Continuous
- Fenestrated
- Discontinuous or sinusoidal

Continuous capillaries are constructed of epithelial cells that overlap tightly, leaving no gaps between them. They are present in the skin, muscles, and lungs as well as the central nervous system (the brain and spinal cord), and are the least permeable type of capillary. Only those substances with **molecular weights** of less than 10,000 can easily cross them. They are particularly important in what is known as the blood-brain barrier. This barrier prevents damaging substances from being transmitted from the circulating blood to brain tissue and to the watery cerebrospinal fluid that cushions and protects the brain and spinal cord. Although continuous capillaries generally permit only small molecules to traverse them, even large molecules with molecular weights of up to 70,000 can make their way across, given enough time. Scientists believe this is accomplished through temporary openings that may occasionally form between epithelial cells.

Unlike continuous capillaries, **fenestrated capillaries** are always full of holes. The endothelial cells overlap much less tightly, creating gaps. In addition, they have numerous pores, or fenestra (literally "little windows"), of 2.0×10^{-6}–3.9×10^{-6} inches (50–100 nm) in diameter. These openings greatly increase the capillaries' permeability, making them particularly useful in tissues that exchange a great deal of fluid and metabolites with the blood. Fenestrated capillaries are common in such tissues as the kidney and the **intestinal villi**, which are tiny projections that serve as the nutrient exchange point for the intestines.

Discontinuous or **sinusoidal capillaries** are large capillaries with apertures so wide that bulky proteins and even red blood cells can pass through them. Of the three types of capillaries, they are the most permeable to water and solutes. They are found in such tissues as the liver, spleen, and bone marrow.

Exchange Function

The exchange of water, oxygen, nutrients, hormones, drugs, waste products, and other chemicals occurs primarily at the capillaries. Oxygen and carbon dioxide, which are gases, move by passive molecular flow, a process called **diffusion**, right through the wall. An individual endothelial cell, like other cells in the human body, is surrounded by a thin membrane made of two fatty layers, the lipid bilayer. Substances that can pass through this layer are termed lipophilic (fat-soluble). Besides oxygen and carbon dioxide, other substances that can readily cross the membrane include some drugs, like the general anesthetic a person receives before surgery. Actually, oxygen and other materials don't move directly from capillary to cell or vice versa. Instead, they first enter a region just outside the cell. This fluid-filled extracellular area is called the **interstitial space**.

The direction of diffusion is determined by the **concentration gradient**, which means that molecules travel from an area of high concentration to one of low concentration. Transport from high to low concentration is described as moving down the concentration gradient. Therefore, if an arteriole delivers oxygen-laden blood to a capillary near oxygen-poor tissue, the oxygen (O_2) will pass from the blood to the tissue, or from an area of high concentration to an area of low concentration. The same thing happens with the waste product carbon dioxide (CO_2), which accumulates in tissue. The carbon dioxide moves from the cell to the blood, or from an area of high concentration to an area of low concentration. The reverse likewise occurs in the lungs, when oxygen-poor blood arrives to pick up oxygen from and drop off carbon dioxide to the **alveoli**, the tiny air sacs in the lungs. In this case, the alveoli gather oxygen with every breath a person takes, then deliver it to the blood in the pulmonary circulation. As mentioned Chapter 2, oxygen in the blood is carried by the large hemoglobin molecule, also known as a **respiratory pigment**. Each hemoglobin molecule can carry four oxygen molecules. Because red blood cells in the average adult number from 24.8 to 28.6 trillion, and each red blood cell contains about 300 million molecules of hemoglobin, the potential for oxygen transport by the blood is immense. (See the Respiratory System volume in this series for more information on this process.)

The oxygen exchange between the capillaries and the alveoli, as well as other tissue cells, is possible because of their proximity to one another. The time it takes for a molecule to move is proportional to the square of the distance moved. The time quickly escalates as oxygen has farther to go. For this arrangement to work in the human body, an enormous number of capillaries is required—enough to place capillaries within 3.9×10^{-4}–7.9×10^{-4} inches (10–20 μm) of just about every cell—just one, two, or three cells' diameter away. Some organs even have special networks or clusters of capillaries. In the kidney, the cluster is known as the glomerulus and facilitates the considerable volume of blood that flows to and from this organ. The capillaries in the glomerulus are typically fenestrated. (See Chapter 7 as well as the Urinary System volume of this series for more details.) Other tissues and organs that have a higher density of capillaries include the heart and skeletal muscles, which are both very metabolically active and demand a highly proficient transfer of gases and nutrients. In organs and tissues, like joint cartilage, that are less active and less oxygen-demanding, fewer capillaries are necessary.

Of course, the body cannot rely solely on diffusion to get oxygen all the way from the lungs to every tissue cell. The pumping of the heart does much of the transportation, forcing the blood along the arteries, then more slowly into the arterioles, and, slower yet, into the capillaries (see photo). Blood is progressing so slowly in the capillaries that each red blood cell commonly takes 1–2 seconds to traverse it, ample time for diffusion to occur.

In smaller blood vessels, like the capillaries, even the tiny red blood cells have to bend, twist, and otherwise deform to squeeze through the very tight passageway. © Dr. David Phillips/Visuals Unlimited.

Large molecules and hydrophobic (fat-insoluble) substances require a bit more than simple diffusion to get the job done. Consider **glucose** ($C_6H_{12}O_6$), a sugar molecule made up of six atoms of carbon (C_6), twelve atoms of hydrogen (H_{12}), and six atoms of oxygen (O_6). In this case, the protein insulin aids in the transport of this rather hefty molecule, which is a key energy source for cells. The insulin, in essence, opens the way for glucose to enter cells. Arterial blood already contains a higher concentration of glucose, so once the cells become receptive with insulin's help, diffusion can occur. In fact, glucose can traverse a capillary wall in just 0.0005 seconds, then enter an adjacent tissue cell just 0.05 seconds later.

As described, fenestrated capillaries have numerous pores to accommodate the movement of water, hydrophilic nutrients, and other materials. The discontinuous, or sinusoidal, capillaries allow the transport of very large molecules, even red blood cells. Water, on the other hand, flows from blood to tissues and back on a pressure gradient rather than a concentration gradient. This involves the lymphatic system, which will be discussed in Chapter 7.

Before leaving this section, it is important to note that while capillaries are the primary vessels for exchange between the blood and tissues, some small arterioles and venules that lie adjacent to capillaries may also participate in this process. These are known as the **metarterioles** and the **postcapillary** or **pericytic venules**. Despite their contributions, the capillaries remain by far the foremost exchange site.

Measuring Blood Pressure

Medical professionals use a blood-pressure cuff, or sphygmomanometer, to measure blood pressure. It is an inflatable piece of material that is wrapped around the patient's arm just above the elbow.

When pumped up with air, the device temporarily disrupts blood flow to the lower arm. A medical professional then listens to vascular movement through a stethoscope and watches a pressure monitor as the air is released from the cuff. The point at which blood begins to force its way past the compressing cuff is called the systolic pressure. This number, which represents the pressure when the heart beats, is typically 120 mmHg. As compression in the cuff continues to drop, the blood pressure eventually falls below the cuff's compression. This point is called the diastolic pressure, and is typically 80 mmHg. A typical blood pressure is, therefore, 120/80 mmHg.

When blood pressure reaches 140/90 mmHg or beyond, the patient is diagnosed with high blood pressure, or hypertension. A study in the November 2001 issue of *The New England Journal of Medicine* indicated that even systolic blood pressures of 130–139 or diastolic of 85–89 may be cause for concern. They pored through data collected on thousands of people and found increased tendency toward strokes, heart attacks, and heart failure among individuals with blood pressures within those ranges.

4

The Blood Circuit

Now that the basic components and functions of the arterial, venous, and capillary systems have been described, this chapter will take a closer look at how the blood makes the transit through those vessels.

ONE-WAY VALVES

Blood valves are located throughout the circulatory system. Just as a door marked "push" or "pull" only opens in one direction, these valves swing strictly one way. In the arterial system, blood rushes from the left side of the heart through a valve and into the aorta. (Heart contraction and heart valves will be discussed further in Chapter 5.) As the heart beats, the valve swings wide to let the blood pass into the aorta. The blood flow naturally slows when the contraction ends, and the valve swings shut. This prevents the blood from streaming back into the heart chamber.

Other large blood vessels have similar valves that open in just one direction. In the legs, for example, a pair of **semilunar valves** swing outward to the vessel walls, allowing blood to move past. Any reverse flow causes the valves to close. Just as wind from outdoors may cause a door to swing open, but a breeze from indoors can quickly slam it shut, the valves allow blood to move one way, but not the other. This simple system ensures that deoxygenated and waste-filled venous blood doesn't flow backward and mix with the oxygenated, nutrient-filled arterial blood.

BLOOD PRESSURE

Systemic Circulation

Blood pressure is a key force driving the blood through the arterial system. Blood leaves the heart in the systemic circulation under a very high pressure caused by the heart's contraction. On average, the blood pressure in the aorta reaches 120 mmHg following a heartbeat, then falls back down to 80 mmHg before the next heartbeat. This is often written as 120/80 mmHg.

The 120 mmHg reading in the aorta immediately following a heartbeat is the highest pressure that blood reaches in the circulatory system. As the heart's contraction ends, the blood pressure quickly drops. As seen earlier, the aorta and other large arteries have a high percentage of elastin, the protein that permits stretching. When the pulse of blood enters the aorta, that large vessel quickly distends, then slowly returns to its normal size. By doing so, it eases the blood pressure. Subsequent arteries do the same thing, although they become less and less elastic as blood moves farther from the heart. Every time a vessel widens, the pressure drops a bit. By the time the blood reaches the junction between the small arteries and arterioles, it has diminished to about 60–70 mmHg. At the arteriole-capillary border, the blood pressure is just 35 mmHg.

Pulmonary Circulation

The pulmonary circulation (see color insert) begins with the beat of the right side of the heart, which is smaller and less powerful than the left side that drives the systemic circulation. Here, the pressure of the blood leaving the heart and entering the pulmonary artery is just 25 mmHg, which is still sufficient to force the blood the small distance from the heart to the lungs. Between heartbeats, the pressure drops to about 10 mmHg. As in the systemic circulation, the pressure continues to decline as the blood travels from arteries to arterioles to capillaries.

THE RETURN TRIP

If the arterial system requires the force of the heartbeat to drive blood to the lungs and body tissues, what does the venous system use to propel the blood back to the heart? The answer has several parts.

As shown in the section on the venous system, the veins and venules greatly outnumber the arteries and arterioles. They are also oriented in the opposite direction with blood flowing from the smallest and most plentiful vessels into ever-larger but fewer vessels. In the arterial system, the total cross-section of the vessels (the sum of their lumens) increases as they get farther from the heart, which results in a slower blood velocity. This reliance of the flow rate on cross-sectional area is illustrated in the mathematical equation:

$$\text{velocity of the blood (mm/sec)} = \frac{\text{blood flow (mm}^3\text{/second)}}{\text{cross-sectional area (mm}^2\text{)}}$$

The velocity is the overall speed of the blood through a vessel, the blood flow is the volume of blood that moves through a vessel, and the cross-sectional area is the size of the vessel's lumen. The mathematical formula shows that velocity is inversely proportional to the cross-sectional area, so as cross-sectional area increases, the velocity decreases.

In the venous system, the vessels increase in size closer to the heart, but their number decreases dramatically, until the two large venae cavae are accepting blood from thousands of vessels throughout the entire venous system. The sum total of cross-sectional areas of smaller vessels greatly surpasses the area of larger vessels, which results in an increase in blood velocity. In summary, both the vessels of the arterial and venous systems are largest and least numerous near the heart, and smaller and more numerous away from the heart. The difference is in the direction of flow. On the arterial side, blood is moving away from the heart and slows as it goes. On the venous side, blood is returning to the heart and speeds up as it approaches the heart.

In addition, venous blood gets a little help in its return trip from the structure of the vessels, from muscles, and even from arteries. By the time blood makes its way through the arterial system and the capillaries, its pressure in the venules is just 15 mmHg and is nearly nonexistent by the time it reaches the venae cavae. With such little impetus to return to the heart, gravity would influence the blood to pool wherever the body is closest to the ground. This usually doesn't present a debilitating problem because venous vessels have the ability to deflate and reinflate with very little pressure applied to them, so even the small 15 mmHg gradient from venules to heart is sufficient to urge the blood along. The arteries also help. Because arteries and veins are usually in close proximity, the strong pulses of blood that move from the heart and down the arteries put some pressure on the adjacent veins and help circulate the blood. In addition, the venous blood near the heart gets an added incentive from the heart itself. As the heart's valves open to allow in venous blood, suction results and actually draws the approaching blood into the waiting chamber.

The body's skeletal muscles also help by contracting and relaxing, and pressing on venous vessels. This action squeezes the blood back to the heart through the reinflated veins. This type of muscle action produces obvious results in an exercising person, but even when a person is quietly standing, these muscles are continually contracting and relaxing, and pushing blood into veins. Valves in the larger veins assist as well. Once the blood makes its way partially up the leg, the valves prevent it from rushing back down. In the large leg veins, valves occur about every half inch (1.25 cm) along the vessel.

The legs aren't the only parts of the body that have to contend with the effects of gravity. When a person stands up, about 16.9 ounces (500 ml) of blood shifts to the lower legs, so some pooling does occur. The volume of liquid lowers as plasma from the blood filters out of the vessels and into adjacent tissues. This causes an overall drop in blood volume, and the body responds by lowering cardiac output. As a result, less blood reaches other parts of the body, including the brain. Usually the body responds by quickening the heart rate, constricting the venous vessels in the legs, or using nervous control to decrease the amount of blood reaching the legs. When these steps aren't enough, a person begins to feel light-headed. By putting the head between the knees—or, in extreme examples, by fainting—the head is lowered, making it much easier for the blood to fight gravity and reach the head.

EFFECT OF VELOCITY ON CAPILLARIES

The blood's flow rate is also an important consideration in the blood-to-tissue exchange that occurs in the capillaries. Because of capillaries' distance from the heart and the distensibility of the arterial vessels, as well as the large cross-sectional area of the capillaries, the blood's velocity is so slow that blood cells barely creep through the capillaries. In fact, the blood's velocity in the capillaries is less than 1/200th its speed in the aorta. The sluggishness provides ample time and optimal conditions for diffusion of oxygen and other materials to occur. In most cases, blood cells take 1–2 seconds to traverse a capillary, a considerable time for a vessel that is just 0.02–0.039 inches (about 0.5–1 mm) long.

CONTROL OF BLOOD FLOW

Although the circulatory system is similar to a road system in some regards with the vessels analogous to the highways and streets, the cardiovascular system is hardly passive. The blood vessels are alive and dynamic structures that can change diameter and alter the flow of the blood within them.

As previously shown, the blood vessels contain smooth muscle. When contracted, these muscle cells can narrow blood vessels. When the contraction ends, the blood vessel returns to its larger diameter, which is driven by the pressure of the passing blood. Even when the smooth muscle cells aren't contracting, however, they are imparting muscle tone to the vessels.

The mechanism for the control of this tone is called the **Bayliss myogenic response**. Without this response to counteract the force of the blood pressure, the vessels would continue to stretch wider and wider, which would affect flow.

Besides those in the peripheral venous system, other blood vessels change diameter to meet the needs of the body. These types of adjustments occur continually. For example, the brain requires a constant, sufficient blood flow. Oversight and maintenance of this flow is the job of the **arterial baroreceptor reflex**, which responds to slight changes in blood pressure. In this case, the **baroreceptors**, which are pressure detectors located in the major arteries, sense a dip or spike in blood pressure. When pressure increases, the baroreceptors reflexively send a message that shuts down the vasoconstrictor center in the brain's medulla, which is responsible for constricting blood vessels. With the vasoconstrictor center off-line, the blood vessels dilate and pressure drops. At the same time, cardiac output changes to meet demands, and venous vessels likewise dilate. These actions collectively return blood pressure to normal. If, however, the blood pressure remains elevated, as occurs in a person with **hypertension (high blood pressure)**, the baroreceptors become accustomed to the new, higher average pressure and stop sending messages to the vasoconstrictor center. As a result, the body no longer tries to rectify the higher blood pressure. (For more information on hypertension, see Chapter 10.)

The arterioles play a vital role in controlling blood flow. These vessels are often called resistance vessels because they offer resistance to and therefore regulate the flow of blood from the heart. Due to the considerable smooth muscle in the walls of these vessels, the lumen can change diameter to allow more or less blood to reach the tissues. In addition, the arterioles that feed capillaries, called **terminal arterioles**, take their cues from local metabolic factors rather than nerves. Each terminal arteriole serves as a door to its own little network of capillaries. Based on metabolic needs, the arteriole can close tight or open wide to regulate blood flow.

Venous muscle tone is also important in peripheral venous vessels, which regulate blood volume by adjusting the amount of liquid in the plasma and in such tissues as the skin, muscles, and kidneys. These peripheral vessels can thus act as blood reservoirs that can temporarily hold unneeded blood, which has a regulating effect on the overall blood supply.

Hormones regulate blood flow, too. Adrenaline is perhaps the most well-known hormone in this regard. When a person is under physical or mental stress, the body boosts adrenaline production. This hormone, as well as angiotensin II and vasopressin, helps to maintain a healthy blood pressure even under the most dangerous conditions, like severe blood loss. (Hormones are discussed further in the Endocrine System volume of this series.)

In addition to controlling blood flow through muscles, hormones, and nerves, the body has failsafes in certain organs, including the brain. Here, arteries and arterioles merge to create what are called **arterial anastomoses**. These are alternative sites where the organ can obtain blood if a supplying artery becomes blocked. The anastomoses essentially serve as back-up blood sources for use in emergencies. (The arterial anastomoses are not to be confused with arteriovenous anastomoses, which are wide vessels found especially in the skin and nasal mucosa. These vessels bypass capillaries and attach arterioles and venules directly.)

HEAT EXCHANGE

Beyond the transport of oxygen, carbon dioxide, and nutrients, the blood also transports heat. Even when a person is resting, heat conservation is important. The blood in the arteries is typically warmer than that in the veins, so heat transfers from the major arterial vessels to the adjacent main venous vessels—another advantage to having the two systems' vessels next to one another. This helps to conserve body heat.

When a person is active, muscle tissues can become quite warm. In that case, the circulatory system's job is to eliminate some of that heat rather than conserve it. Here, plasma in the blood takes up heat from the tissues, then transfers it to the environment as the warm blood traverses vessels near the skin. The plasma, which is mostly water, acts as the body's "radiator," circulating water to keep the body cool. In addition, capillaries deliver heat from blood to tissues, causing the flushed coloration typical of a person during a physical workout.

The same type of thermoregulation occurs in response to external temperatures. A hot day can cause an increase in core temperature. Sweating, and the evaporation of that sweat, reduces skin temperature. Blood flow in the skin increases and heat is lost to the environment. When a person experiences cold air, on the other hand, skin temperature drops, blood flow to the skin and extremities decreases, sometimes the muscles "shiver" to produce heat, and core heat rises.

5

The Heart: A Living Pump

Although it is only as big as a fist, the heart drives the circulatory system. This organ works constantly over the course of a lifetime, pumping blood to the lungs in the pulmonary circulation and to all other body tissues in the systemic circulation. Even a short pause in its functioning can result in death. This chapter will take a closer look at this amazing organ and how it carries out its job.

ANATOMY AND BLOOD FLOW

The heart is truly a pump made of muscle tissue. Blood moves in and out of this pump through four chambers inside. The two, smaller, top chambers are called atria (atrium is the singular form) and two bottom chambers are called ventricles (see photo and Figure 5.1) It actually works like a double pump with the right half taking in blood returning to the heart from the body tissues, and then sending it over to the lungs to drop off carbon dioxide and pick up oxygen. The newly oxygenated blood then heads back to the other side of the heart, where it gets the added boost it needs to propel it to the body tissues again. This double pump springs into action an average of 70–80 times every minute, all day and all night for a person's entire lifespan.

CARDIAC MUSCLE

The heart is made out of cardiac muscle (also known as myocardium), a tissue that is unlike the smooth or striated muscle seen elsewhere in the human body. Striated muscle is the tissue that a person uses to move his or

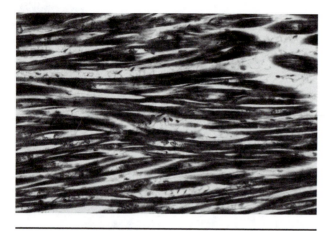

Cardiac muscle. The heart has muscle tissue unlike that found in other areas of the body. These cardiac muscle cells interconnect, which allows the heart to beat as a unit. © M. Peres/Custom Medical Stock Photo.

her legs or fingers. Because the individual can control it, it is also known as **voluntary muscle**. This tissue has light and dark bands, called striations, which give skeletal muscle yet another name: striated muscle. Smooth muscle, like that in blood vessels, is known as **involuntary muscle**, because a person can't direct its movements like he or she can control skeletal muscle. Instead, the autonomic nervous system controls its action. Falling somewhere in the middle of these two types of tissue is cardiac muscle. Cardiac muscle has

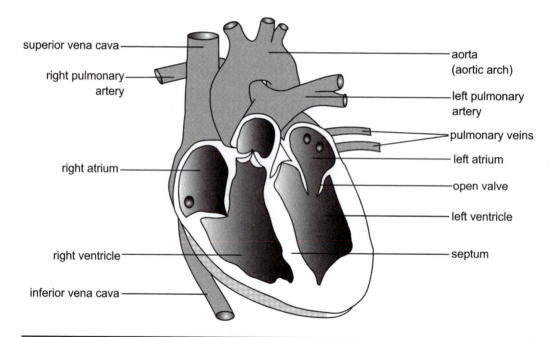

Figure 5.1. A simplified illustration of the heart with main features shown.
The heart has two upper chambers, called the atria, and two lower chambers, or ventricles. The right and left sides of the heart are separated by a thick wall, known as the septum, and the atrium and ventricle on each side are divided by valves that open to allow the passage of blood. The superior and inferior vena cavae are the main blood vessels that return deoxygenated blood from the upper and lower body, respectively, to the heart's right atrium. The right and left pulmonary arteries deliver blood from the right ventricle to the lungs for oxygenation, and the newly oxygenated blood returns to the left atrium through pulmonary veins. Newly oxygenated blood leaves the heart by way of the aorta, the largest artery in the body. Near the heart's right ventricle, the aorta bends. This is known as the aortic arch.

the striations seen in skeletal muscle, but it takes its direction from the autonomic nervous system like the smooth muscle does. Unlike either striated (also known as skeletal) or smooth muscle, cardiac muscle cells are very closely linked to one another and have fibers that interconnect one cell to the next. As will be shown in the section on electrical activity later in this chapter, this is vital in making the heart beat as a unit. In addition, cardiac muscle doesn't tire out like skeletal muscle does, and it requires a shorter resting time between contractions. It's easy to assume that skeletal muscle can contract for a very long time, especially when considering how a body maintains muscle tone. A closer look reveals that different groups of skeletal muscle alternately shorten to give the appearance of constant contraction, even when the muscle cells are individually contracting and relaxing. In the heart, conversely, all of the cardiac cells contract at the same time. (For an in-depth discussion of cardiac muscle, see the volume on the Musculoskeletal System in this series.)

This muscle tissue of the heart surrounds all four of its chambers, completely enclosing them. The muscular walls of the atria don't require the kind of force that the ventricles do, and they are considerably thinner. All of the chambers are lined with endocardium, a thin membrane that provides a smooth, slick surface for the blood to slide along. This membrane is similar to the endothelium that coats the inside of blood vessels. The entire heart is enclosed in a fluid-filled fibrous sac, called the **pericardium**, that attaches to the diaphragm. Actually a double sac with fluid between the two layers, the pericardium supports, lubricates, and cushions the heart. The diaphragm, which is a large muscle that separates the chest and abdominal regions, pulls downward during inhalation and consequently tugs the heart into a more upright position.

As mentioned, the atria sit at the top of the heart and the ventricles at the bottom. The word atrium means "entrance hall" in Latin. These chambers got their name because the blood enters the heart through the atria. At one time, atria were known as auricles because they somewhat resemble ears, and *auris* is Latin for ear. Ventricle is Latin for "little belly," and one might say they look like ministomachs. If the atria are considered the entrances

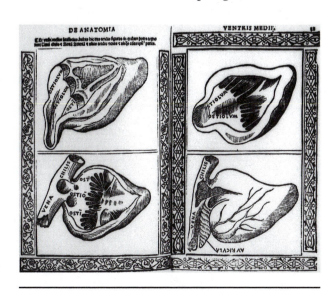

Four views of the heart. Woodcut by Jacopo Berengario do Capri, from *Isagogae breves* . . . in *anatomia humani corporis* . . . , 1593. © National Library of Medicine.

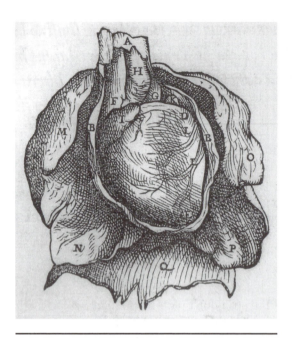

Andreas Vesalius, "The pericardium surrounding the heart." From *De humani corporis fabrica.* © National Library of Medicine.

to the heart, the ventricles are the exits, and blood drains from them via the pulmonary artery and aorta. The right and left sides of the heart are completely separated by a thick muscular wall called the septum. This ensures that deoxygenated and oxygenated blood don't mix.

In addition, the heart has a ring of fibrous connective tissue, called the **annulus fibrosus**, that serves as an anchor for the heart muscle and as an almost-continuous electrical barrier between the atria and ventricles. Electrical charge plays a key part in proper heart functioning, and will be discussed later in this chapter.

HEART VALVES

The right atrium is separated from the right ventricle by a valve. The same holds true for the left atrium and left ventricle. This arrangement allows blood to move from one to the other, but only when the valve is open. Like the valves in the blood vessels, the heart valves permit blood flow in only one direction. Between the two right chambers, the valve has three flaps, or cusps, and is known as a **tricuspid atrioventricular valve**. It is commonly called the **AV valve**. A **bicuspid** (two-cusped) **mitral valve** (also known as the **mitral valve**) divides the two left chambers. The ventricles have additional valves to seal off flow between them and the arteries. The valve between the right ventricle and pulmonary artery is called the **pulmonary semilunar valve**, and the valve between the left ventricle and aorta is called the **aortic semilunar valve**. Both have three cusps.

The AV, mitral, and semilunar valves all work the same way. When blood is rushing through a valve, each of which is about 0.004 inches (0.1 mm) thick and made of fibrous connective tissue lined with membranous tissue, its cusps open with the flow. As the blood tapers off, the cusps fall back to their original closed positions. Tiny tendinous cords, appropriately called **chordae tendineae**, attach to adjacent muscles (papillary muscles) and prevent the valve's cusps from falling back too far and letting blood seep through.

BLOOD FLOW

The valves, chambers, and muscle tissue are all involved in blood flow through the heart (see color insert). Venous blood returning from the body tissues enters the heart through the two large venae cavae. The anterior vena cava carries blood from the head, neck, and arms, and the posterior vena cava conveys the blood from the rest of the body tissues (except the lungs). The right atrium fills with blood from the venae cavae, as well as the coronary sinus, which is the main vein carrying blood from the heart (details of the coronary circulation are included in the next section). When it contracts, the pressure of the blood builds, forcing open the tricuspid AV valve and allowing the blood to flood the right ventricle. When the atrium relaxes, the blood's pressure drops, and the AV valve falls back to its original position. The now-filled ventricle contracts. The only way out for the blood is through the pulmonary semilunar valve, which opens outward into the pulmonary artery.

Blood rushes out of the heart and into the pulmonary artery, where it branches and eventually reaches the lungs and alveoli to pick up oxygen. Following the ventricle's contraction, the blood pressure in the pulmonary artery is 35 mmHg. The newly oxygenated blood returns to the left side of the heart and enters the atrium. It contracts, forcing blood through the bicuspid mitral valve in the left ventricle. Now that the blood is in the left ventricle, the mitral valve closes, the ventricle contracts, and blood flows through the aortic semilunar valve and into the aorta. The force of the contraction boosts the blood pressure in the aorta to 120 mmHg, more than three times the pressure of a similar volume of blood in the pulmonary artery. To accomplish this feat, the muscular wall of the left ventricle is about three to four times as thick as the right ventricle's wall. The increased thickness gives the left ventricle the power it needs to drive blood throughout the body. Normally, the amount of blood in a completely filled adult ventricle is about 0.12 quarts (120 ml). This amount is called the **end-diastolic volume**. The heart typically only ejects about two-thirds of this blood, retaining 0.04–0.05 quarts (40–50 ml). That yields a total stroke volume—the amount of blood that exits the heart—of 0.07–0.08 quarts (70–80 ml). When the body needs to increase stroke volume, such as during periods of heavy exercise, it calls on the heart to begin pumping the residual 0.034 quarts (40–50 ml).

As it turns out, both atria fill simultaneously, so the right side of the heart is taking in deoxygenated blood from the systemic circulation at the same time that the left side is admitting newly oxygenated blood from the pulmonary circulation. The ventricles likewise fill at the same time. When either the ventricles or the atria contract, the chambers get smaller. When the atria relax following a contraction, they go back to their normal, larger size. The ventricles have an effect on the size of the atria, because they contract when the atria are resting. As the ventricles contract or shorten, they actually pull down on the bottom of the atria to further expand these upper chambers. This enlargement creates suction in the atria and serves to draw in blood from the venae cavae and begin preparing the way for the next heartbeat.

The heart spends about as much time resting as it does contracting, with each contractile and resting period lasting less than a half of a second. The contractile period is known as **systole**, and the resting period as **diastole**. Although each contraction lasts only about 0.4 seconds, that is enough time for the atria and the ventricles to contract. Careful observation reveals that the atria contract slightly before the ventricles during systole, so that blood travels from the atrium to the ventricle and to the aorta or pulmonary artery with every contraction.

The heartbeat's familiar lubb-dupp sound is actually the valves vibrating when they flap shut rather than the heart muscles expanding and contracting as many people think. These vibrations (about 100 cycles per second, or 100 Hertz) result when the heart's contraction ends and the blood starts to slosh back toward the valves. As the liquid hits the valves, they shudder slightly. That shudder is the vibration that is audible to a doctor with a stethoscope pressed to a patient's chest. The "lubb" half of the heartbeat is the closing of the tricuspid AV and mitral valves that separate the atria and ventricles. The "dupp" is the sound made by the semilunar valves located between the ventricles and either the aorta or pulmonary artery.

CORONARY CIRCULATION

The heart is a hardworking muscle that demands its own arteries and veins to maintain its operation. In fact, some textbooks and medical articles even describe the human body as having three circulatory systems: the systemic, the pulmonary, and the coronary circulations.

The two major arteries feeding the entire heart muscle are the right and left coronary arteries that stem from the base of the aorta. The right coronary artery primarily delivers oxygen-rich blood to the two right chambers: the right atrium and the right ventricle. The left coronary artery mostly feeds the left side. Unlike the right artery that remains as a single, large vessel, the left almost immediately splits into two vessels, known as the transverse and descending branches. Each artery also ships a little blood to the opposite side, but the right artery mainly concentrates on the right side, and the left artery on the left side. Strangely, the amount of blood delivered by the two sides differs in individuals. About half of all people have a dominant right artery, three in ten have equal delivery in the two arteries, and about one out of five have a dominant left artery.

As mentioned, the heart demands a strong blood flow to supply the oxygen it requires. The body will even respond to a blockage in a coronary artery by rerouting blood through nearby **collateral arteries** and around the compromised area. Heart patients frequently refer to this phenomenon as "growing new arteries." In addition, the heart has more than 2,000 capillaries per 0.00006 inch (mm^3) that help ensure an ample oxygen supply to this active muscle.

ELECTRICAL ACTIVITY

How does the heart keep pumping? What triggers a heart to beat? Electrical activity is the answer. In the human body, the nervous system controls the overall electrical activity, including that in the heart, and is thus its primary regulator.

In skeletal muscle, nerves outside the muscle direct its contraction. In the heart, small and weakly contractile modified muscle cells serve as the initiation point for the heart's electrical system. These cells, located in a 0.8 × 0.08 inch (2 cm × 2 mm) area in the atrial wall near the entrance of the superior vena cava, are collectively known as the **sinoatrial node (SA node)** or the **pacemaker**. These cells are electrically connected, so when one "fires"—delivers an electrical impulse—they all do. The pacemaker fires spontaneously and needs no nervous system input to continue to deliver regular electrical impulses. In fact, the heart will continue to beat for a while even if it is completely removed from the body. On the other hand, the nervous system is important in that it can override the pacemaker's regular firing rate and either slow it down or quicken it. The heart also has back-up regions that can take over if the pacemaker is compromised. These regions, called **ectopic pacemakers**, are capable of initiating the heartbeat when necessary. (A more detailed description of ectopic pacemakers is included in Chapter 10.)

When the pacemaker fires, the electrical impulse spreads at a rate of about 3.29 feet (1 m) per second to the left and right atria and causes them to contract. As noted earlier, the fibers in each cardiac muscle cell are connected to fibers in adjoining cells. This connection allows the cells to contract nearly in unison. As the atria contract, the blood flows past the respective valves and into the left or right ventricle. At the bottom of the septum dividing the atria is a small group of cells and connective tissue known as the **atrioventricular node**, or **AV node**. The AV node is the only conducting path through the annulus fibrosus that divides the atria and ventricles. This node gets the electrical impulse from the pacemaker at about the same time as the atria do, but forwards it much more slowly to the ventricle, resulting in a delay of about a tenth of a second. This allows time for the atria to contract and squeeze the blood into the ventricles. The AV node then relays the impulse not directly to the ventricle, but to two structures. The first of the two is the **bundle of His** (pronounced "hiss"), a thick conductive tract that transmits the signal to a mesh of modified muscle fibers, called the **Purkinje fibers** (pronounced purr-kin-gee), in the base of the ventricle wall. It is these fibers that pass the impulse to the ventricle (at a speed of 5.5 feet [1.6 meters] per second!). The ventricle contracts beginning at the bottom. As the wave of contraction progresses upward, it efficiently forces the blood upward to the exit valves.

An **electrocardiograph** (**ECG** or **EKG**) records this electrical activity as a jagged line on a sheet of paper (see Figure 5.2). The resulting sheet is called an **electrocardiogram**, which likewise goes by the initials ECG. Each electrocardiogram is divided into five parts that are signified with the letters P, Q, R, S, or T. "P," known as a **P-wave**, is a small bump in the baseline. It indicates an electrical impulse passing from the pacemaker to the AV node. "Q" comes next as a small dip below the baseline, and notes the passing of

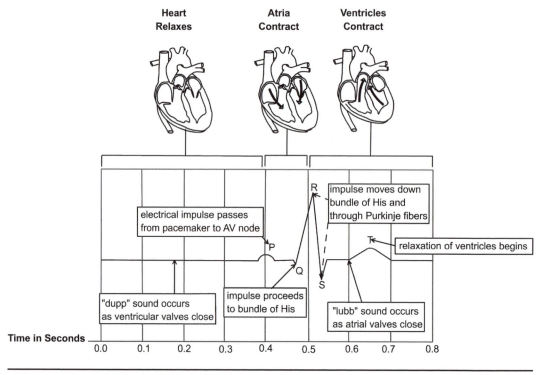

Figure 5.2. An electrocardiogram (ECG) of one cycle of a working heart.
The heart needs less than a second to complete a cycle. The heart's electrical activity is depicted on an ECG as a jagged line, each "jag" noted with a letter that corresponds to a certain aspect of the heartbeat. The "lubb-dupp" sound of a heartbeat occurs as valves in the heart swing shut.

the impulse along the bundle of His. "R" spikes upward to the highest point in an ECG, and "S" follows immediately with a dive below the baseline to the lowest point on an ECG. Both R and S reveal the electrical impulse's movement down the bundle and through the Purkinje fibers. The last phase, "T," is a low spike that gradually returns to the baseline. T represents the relaxation of the ventricles. The entire cycle—all five phases—takes about 0.8 seconds. The "lubb-dupp" of the heartbeat occurs when the heart valves swing shut. The "lubb" occurs between the S and T phases when the atrial valves close, and the "dupp" occurs between the T and P spikes when the ventricular valves close.

NORMAL HEART FUNCTION VARIATIONS

Although the heart beats about 70–80 times per minute in a resting adult, that rate can vary considerably depending on whether the person is sitting down or standing up, walking or running, relaxed or under stress—activity of nearly any sort, as well as various medical conditions and medications, can cause the heart to speed up or slow down. In addition, stroke volume (the amount ejected from the left ventricle to the aorta with each beat) can also change to meet demands.

This formula is used to obtain the cardiac output, which is the volume of blood pumped over a certain period of time (typically the amount of blood ejected by one ventricle in one minute):

$$(\text{stroke volume}) \times (\text{heart rate}) = \text{cardiac output}$$

On average, the stroke volume of an adult at rest is about 0.08 quarts (75 ml). When multiplied by the average male's heart rate of seventy beats per minute (a female's is about seventy-eight beats per minute), the cardiac output comes to 5.5 quarts (5.2 liters) per minute. Cardiac output varies greatly among individuals, and while 5.5 quarts per minute is the approximate overall average, a healthy individual can fall within the range of about 4.2–7.4 quarts (4–7 liters) per minute. Generally, an individual's cardiac output at rest is 3.2 quarts (3 liters) per minute for every 3.3 square feet (or 1 square meter) of body surface area. Using that calculation, an adult weighing 150 pounds (68 kg) and having a body surface area of about 5.9 square feet (1.8 square meters) would have a cardiac output of 5.4 quarts (5.1 liters) a minute.

That cardiac output is distributed to the body more or less on the following principle: the higher the metabolic rate of the tissue, the higher percentage of the cardiac output it receives. Muscles use about a fifth of the oxygen a person breathes, so they receive about a fifth of the blood. Two of the most notable exceptions to the rule are the kidneys and the heart. The kidneys receive as much blood as the muscles, even though the kidneys use less than a third of the oxygen. High blood flow to the kidneys is necessary because these organs are the body's blood filters, removing waste products and excess water. The heart, on the other hand, receives proportionately less blood than it should, if figured according to the rule. It compensates, however, by drawing more oxygen from its limited supply than other organs do, thanks to its particularly dense capillary network.

For the most part, the autonomic nervous system rules the variations of the heart rate, which in turn affects cardiac output. The autonomic system, which controls involuntary activities, has two major divisions: the **sympathetic nervous system** and the **parasympathetic nervous system**. (See the Nervous System volume for more information.) The former stimulates the pacemaker and boosts the heart rate, while the latter inhibits the pacemaker and lowers the heart rate. The sympathetic and parasympathetic systems also have opposite effects on the arteries, with the former causing them to contract and the latter causing them to dilate. In effect, the sympathetic side prepares a person to respond to stressful situations by heightening blood flow, and the parasympathetic (also known as **vagal**) side brings a person back to normal.

Besides the autonomic nervous system, other factors can influence cardiac output, including various hormones, baro- and chemoreceptors, and the fitness level of the individual.

HORMONES

A wide variety of hormones can affect heart performance, but the hormone most commonly associated with heart rate is adrenaline, also known as **epinephrine**. The name adrenaline comes from its point of origin in the adrenal gland that sits on top of the kidney. Adrenaline is the hormone that allows an individual's body to spring into action in the so-called fight-or-flight response: fight for your life, or run for your life. For the cardiovascular system, this hormone causes the heart to race and the blood pressure to soar. It simultaneously constricts the vessels leading to the digestive system and the skin, and dilates those directed to the muscles, heart, and brain. This reroutes blood away from those areas that aren't needed for fight or flight, and to those areas that are required for emergency action.

Usually, adrenaline is secreted in tandem with noradrenaline, or norepinephrine, a hormone that also increases cardiac output. Norepinephrine works by regulating calcium, an element that affects nerve firing in heart cells and other cells. When norepinephrine is secreted, calcium floods the cells, the nerves respond, and the heart contracts more forcefully. This hormone and other factors that enhance contractility are called **inotropic agents**. As mentioned earlier, each beat of the heart typically only ejects about two-thirds of the blood, retaining 0.04–0.05 quarts (40–50 ml) in the heart chambers. When the body demands an increased stroke volume, a more forcefully contracting heart begins to pump that residual volume.

RECEPTORS

Every time the heart beats, the blood pressure increases in the aorta and in the carotid sinuses, which are swellings or expansions at the base of the internal carotid arteries. When the heart's contraction ends, the pressure decreases. Detectors in the walls of major arteries perceive those pressure changes by sensing the tension in the vessel walls. The detectors, neurons called baroreceptors, pass this information to the parasympathetic nervous system, which responds as necessary by lowering the heart rate and decreasing its contractility, which together cause a dip in blood pressure. The sympathetic nervous system also heeds the call by reducing arterial tone and making arteries more elastic, which also serves to decrease blood pressure. The major baroreceptors are the aortic baroreceptors that keep track of blood pressure in the ascending aorta, and the carotid sinus baroreceptors in the neck that track blood flow to the brain. In addition, atrial baroreceptors at the venae cavae and right atrium check blood pressure as blood enters the heart from the venous system. When more blood is entering the heart than is being pumped out, the body rectifies the imbalance by heightening cardiac output until the incoming and outgoing blood are equalized again. In summary, the baroreceptor system is an effective means for evening out the short-term spikes and dips in blood pressure.

Besides pressure detectors, the body has chemoreceptors to monitor the levels of oxygen and carbon dioxide in the blood. These receptors, located near the carotid sinus and the aortic arch, also detect acidity, or pH, levels. The amount of carbon dioxide in the blood can alter its acidity level, because carbon dioxide dissolves in the water of the blood plasma and makes carbonic acid. The more acidic the blood, the lower the pH. When chemoreceptors sense a rise in carbon dioxide levels, a fall in oxygen levels, or a drop in pH, they trigger a hike in cardiac output, which leads to higher arterial blood pressure. Other chemoreceptors in the medulla oblongata (lower brain stem) keep track of blood composition going to the brain and respond to reduced oxygen by initiating the dilation of cerebral vessels while constricting vessels to other parts of the body. The body, then, works to maintain blood flow to the brain at the expense of other organs and systems.

FRANK-STARLING MECHANISM

When additional blood enters the heart from increased venous return, the amount of blood in the ventricle similarly rises. The heart responds by more forcefully contracting to push out that additional blood. Conversely, when a less-than-normal amount of blood enters the heart from the venous system, heart contractility decreases proportionately. The mechanism that regulates this rationed heart response and the related change in stroke volume is called the **Frank-Starling mechanism**, or **Frank-Starling relationship**.

The Frank-Starling mechanism comes into play frequently. Whenever a person shifts from lying down to standing up, for instance, the venous return drops. That decrease in venous return—typically 0.4–0.5 quarts (400–500 milliliters) in an adult—results in a 20 percent drop in cardiac output, as well as a plunge in arterial blood pressure. The decreases evoke a quickening of the heart rate and constriction of the veins in the lower extremities to counter gravity and push the blood back toward the heart.

EXERCISE

Normally, the average adult has a heart rate of about 50–60 beats per minute (bpm) when sleeping, about 70 bpm when awake and relaxing, and 100 bpm or more when under physical or mental stress. Stressful situations also generate a spike in cardiac output. For example, a sudden scare can temporarily boost output by 20 percent or more, and exercise among nonathletes can triple cardiac output. In comparison, an extremely fit individual when exercising may have a heart rate of 50 bpm and a cardiac output nearly twice that of an untrained person. Cardiac output in the best-trained individuals may reach 9.25 gallons (35 liters) per minute, *seven times* the average adult's resting output of 5.5 quarts (5.2 liters) per minute.

When a person exercises, the body boosts blood flow to the skeletal muscle in use as well as to the heart. If the person is doing particularly strenuous exercise, blood flow to these areas is heightened even more. In heavy exercise, more than three-quarters of the cardiac output may go to the muscles in use. As noted earlier, the body accomplishes this shift in blood flow by increasing cardiac output and restricting blood flow to those regions of the body, like the digestive system and kidneys, that aren't immediately required. The brain's blood supply remains stable. The body accomplishes the rise in cardiac output to the muscles and heart primarily by increasing heart rate, but also by boosting contractility and, thus, stroke volume. According to the formula

$$(\text{stroke volume}) \times (\text{heart rate}) = \text{cardiac output}$$

a rise in heart rate causes a similar rise in cardiac output, even when the stroke volume is unchanged, so measuring one's pulse during jogging or some other aerobic exercise is a good indicator of the level of training. The heart has its limits, however, and a heart rate that is too high can so limit the time that the ventricles have to contract that they become unable to pump out the same volume of blood. When this happens, stroke volume declines. As a general rule, the maximum heart rate in an individual is:

$$\text{maximum heart rate per minute} = 220 - \text{age (in years)} \pm 10$$

A 15 year old, then, would have a peak heart rate of 185–205, whereas a 70 year old's top rate would reach 140–160. Overall, stroke volume in an untrained individual may increase 10 percent during exercise, jump 35 percent in a well-trained individual, and nearly double in an Olympic distance runner.

In addition to changes in heart rate, blood flow also increases due to the action of the contracting muscles, which encourage the flow of venous blood back to the heart. When a person begins exercising, blood flow to the skin is also decreased, because the skin is not involved. As a person heats up, however, the blood vessels dilate again and the blood carries body heat to the skin where it is released.

After exercise, the body returns to resting levels quite quickly. The heart rate slows, cardiac output drops, and rate of respiration decreases. The blood vessels in the muscles remain dilated for a while longer, then gradually go back to normal.

Athletes, particularly those who are involved in endurance sports, undergo conditioning to maintain high cardiovascular performance. Continued training promotes the formation of additional capillaries around well-used muscles and in the heart, as well as thicker, more powerful ventricles that surround enlarged ventricular chambers. Bigger chambers hold more blood, and stronger ventricles improve contractility. Well-conditioned athletes, therefore, increase cardiac output by amplifying stroke volume, which is also known as the end-diastolic volume (the amount of blood in the left ventricle just before it contracts).

Even people who aren't world-class athletes, but who exercise regularly, can see cardiovascular results for their efforts. As little as 6–12 months of endurance training can alter the resting heart rate from the 70 beats per minute and 0.08 quarts (70 ml) stroke volume in the average sedentary individual to just 55 bpm and 1 quart (90 ml) per beat in the new exerciser. Notice that the heart beats more than 20 percent slower, but the cardiac output (stroke volume × heart rate) remains the same. Olympic athletes and other heavy trainers can achieve the same cardiac output with heart rates of just 30–40 bpm. The comparison between heart activity in a sedentary individual, casual jogger, and Olympic athlete is akin to the speed of pedaling required to ride an 18-speed bicycle at the same speed in first gear (sedentary individual), ninth gear (casual jogger), and eighteenth gear

(Olympic athlete). A rider on a bicycle in eighteenth gear pedals far fewer times to achieve the same speed as a bike rider in first gear.

Only as big as a fist, but working every minute of every hour for a lifetime, the heart is an amazing organ. In the next chapter, we'll see how the heart and the rest of the cardiovascular system function in mother and fetus.

Growing Tissue That Beats

Researchers have been able to grow tissues from cells for a while, but only recently have they been able to grow heart tissue that actually contracts like a normal heart does.

MIT's Lisa Freed and Gordana Vunjak-Novakovic were able to grow the beating heart tissue using a bioreactor originally designed by the National Aeronautics and Space Administration (NASA) to perform cell studies in space. The two researchers placed about 5 million individual heart cells into the bioreactor with an intricate, biodegradable fiber mesh that served as a temporary scaffold that encouraged the cells to connect and grow.

Within a week in the bioreactor, the cells formed tissues that began to beat spontaneously. They also beat synchronously, meaning that the whole piece of tissue contracts at the same time.

While the work is still in its preliminary stages, it offers a view of a future where heart ailments could be corrected with replacement heart tissue grown on demand.

6

From Fetus to Birth

The transformation from egg to newborn involves amazing growth and spectacular development. From the joining of a single minute egg and a sperm cell, an infant arises in just nine months. Cells of the developing fetus, like other living cells, require oxygen and nutrients, and therefore need a circulating blood supply that has access to those items. Because the fetus is in a contained environment (a mother's womb), it must obtain its oxygen and nutrients from outside. "Outside" is the mother's circulatory system. Everything the mother ingests or inhales, therefore, maybe transferred to the fetus. This is why so many different foods, tobacco products, alcoholic beverages, and drugs carry special warnings directed at expectant mothers.

FETAL HEART DEVELOPMENT

Shortly after conception, a circulatory system begins to develop in the embryo. At twenty-three days, the embryo is not much bigger than a sesame seed and looks more like a tiny, fictional space alien than a human, but it already has a pair of rudimentary blood vessels that are linked together. In just two days, the area of linkage fuses into a single, hollow bulge that is enveloped with a thin layer of muscle. The bulge begins to contort, and in 24–48 hours, it forms a sideways "S" shape, commences pulsing, and although it is still just 0.04 inches (0.1 cm) long, it starts to look more like the heart it will eventually become. Over the next week, upper and lower chambers form, and the upper chamber splits into the right and left atria. At this point, the entire developing fetus is still smaller than a pea.

When the fetus reaches about six weeks old (and is about as big as a new

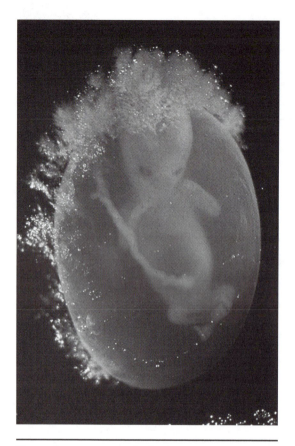

Human fetus at nine weeks. Both the heart and circulatory system are already formed and functioning. © Collection CNRI/Phototake.

pencil eraser), the septum is taking shape and dividing the lower chamber into the right and left ventricles, and valves are developing to maintain a one-way flow of blood. At this point, the fetus grows more quickly, reaching a length of 1 inch (2.5 cm) by eight weeks (see photo). Now, the heart is functioning as a four-chambered, pumping organ with a one-way system that delivers blood throughout the body. The circulatory system works just like it does in an adult—with one very noticeable exception. The fetal circulation is linked to the circulatory system of another human being: the mother.

The development of the heart is a good example of "ontogeny recapitulates phylogeny." Ontogeny is the development of an individual from conception to birth, whereas phylogeny is the evolution of organisms over geologic time. Scientists believe animals evolved in this sequence: fishes to amphibians to reptiles to mammals. The fetal heart development likewise goes through several major steps that follow the evolutionary tree. The one-chambered, tubelike heart resembles that of a fish's heart; the two-chambered heart looks like that of a frog; the three-chambered version like that of many reptiles; and the final, four-chambered version is characteristic of mammals.

THE MOTHER-FETUS LINK

The connection between mother and developing embryo occurs early on. Immediately after conception, the fertilized egg floats unattached inside the Fallopian tubes or uterus of the mother's reproductive system (for a detailed description, see the volume on the Reproductive System in this series) as it develops into a ball of cells called the **blastocyst**. One end of the ball is the **embryocyst**, which will eventually form the fetus. The rest of the ball, called the **trophoblast**, will become the passageway between the fetus and the mother. The entire blastocyst attaches to (or implants on) the uterine lining about a week after fertilization, and begins to accept oxygen and nutrients and rid itself of waste products by diffusion through its contact with the uterine tissues. This arrangement is temporary. Within another three weeks, a more advanced gateway between mother and embryo has formed. This gateway comprises the **umbilical cord** and the **placenta**. The placenta is a rather round and flat structure nearly 8 inches (20 centimeters) wide that links the mother to the umbilical cord, which then attaches to the fetus. While this connection forms, the embryo is also beginning to make blood cells. Unlike adult red-blood-cell formation, or erythropoiesis, which occurs in the bone marrow, fetal erythropoiesis occurs in the liver and spleen. As noted earlier, the embryo's rudimentary heart is already pulsing by the time the mother and embryo have connected.

Through the placenta and umbilical cord, the fetal and maternal circulatory systems are now connected. The mother's blood delivers oxygen and nutrients to the developing fetus, and removes its carbon dioxide and waste products. For the fetus, the blood exchange occurs in capillaries that are located in fetal villi, branching structures in the placenta that are in contact with maternal capillaries. Just as in the adult circulatory system, carbon dioxide and waste products are exchanged for oxygen and nutrients through capillaries. A major difference between the adult and fetal systems is that these capillaries are located in the placenta rather than in the fetus itself. Two **umbilical arteries** and one **umbilical vein** connect the fetus to the capillaries in the villi.

FETAL BLOOD FLOW

Another critical difference between fetal and adult circulation lies in blood flow (see Figure 6.1). In an adult, the newly oxygenated blood flows from the heart to the body through the arterial system, then the depleted, or "used," blood flows from the body back to the heart through the venous system. In a fetus, however, the umbilical vein transports blood that has just picked up oxygen and nutrients from the mother's blood cells, and the umbilical arteries ship depleted blood back to the villi. To deliver oxygen and nutrients to all of its cells, the fetus uses a different circulatory route, and has a few additional structures not seen in adults. These structures include the **ductus venosus** and **ductus arteriosus**, which are defined below, and the **foramen ovale**, an opening to the left atrium.

If the fetal route is viewed as beginning at the capillary exchange site in the placenta, the blood flows like this:

1. Blood from the capillaries flows into the umbilical vein, which makes its way toward the liver.

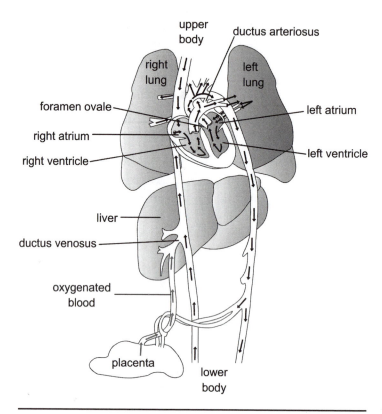

Figure 6.1. Fetal circulation.
With nonfunctional lungs, a fetus receives oxygenated blood from the mother by
way of the placenta. That blood travels up the umbilical vein to the liver. About half
enters the liver, but the rest is shunted to the inferior vena cava by a fetal structure called the ductus venosus. The inferior vena cava carries this blood, along with
blood collected from the lower body to the right atrium, where it joins blood delivered by the superior vena cava that drains the upper body. Some blood moves
from right atrium to right ventricle, but most is diverted through another fetal
shunting structure called the foramen ovale into the left atrium. Blood then flows
into the left ventricle and out through the aorta. A fetus also employs one other
structure, called the ductus arteriosus, which is a shunt between the pulmonary artery and the aorta.

2. About half of the blood enters the liver, then flows into the inferior vena
   cava. The other half runs into the ductus venosus, a shunt that bypasses
   the liver and sends blood directly to the inferior vena cava. The inferior
   vena cava already contains "used" blood that has been collected from the
   lower trunk and extremities, and is returning to the fetal heart.

3. The blood from the inferior vena cava enters the right atrium just as systemic blood does in an adult, but the similarity ends there. Only a small
   fraction of the blood streams into the right ventricle. The rest divides off

via a routing mechanism called the **crista dividens** (the edge of the fora-men ovale), and rushes through an opening between the two atria and into the left atrium, which then empties into the left ventricle.

4. From the left ventricle, the heart pumps the blood through the aorta and into the head, arms, and coronary circulation. After it reaches its destina-tion, the blood transits the superior vena cava and returns to the right atrium, where it mixes with the fraction of blood taking the right atrium–to–right ventricle path. As in adults, the right ventricle empties into the pulmonary artery.

5. Most of the blood in the pulmonary artery bypasses the still-nonfunctioning lungs and instead takes the third, unique conduit through the fetal circu-latory structure. This structure, called the ductus arteriosus, is a shunt that connects to the descending aorta. Once there, the blood bifurcates again with nearly two-thirds leading back to the placenta to begin another cir-cuit, and the remainder flowing out to the lower body in the same manner as in the adult systemic circulatory system.

CHANGES AFTER BIRTH

In a matter of moments, the fetus trades its warm, stable environment and pampered lifestyle inside the mother's womb for the harsh reality of life on its own. In the circulatory system, the shift is dramatic and al-most immediate (see Figure 6.2). Just a few breaths by the newborn are enough to change the circulatory system from one that relies on an ex-ternal source of blood for oxygen to one for an independently function-ing being.

In preparation for this step, the circulatory system had continued to di-rect a bit of blood to the lungs to continue their development. Now pre-pared to spring into action, the lungs expand with the help of chest muscles and the diaphragm to draw in air. Pressure changes in the lungs backtrack to the heart, causing the foramen ovale to close and eventually seal shut after about three months. The umbilical vein, umbilical arteries, ductus venosus, and ductus arteriosus also close. These arteries and veins close almost immediately after birth, while the ductus arteriosus may take up to ten days or more to seal. With the lungs now solely responsible for the body's oxygen, their fetal blood allotment of a small percentage of the pulmonary artery's supply is no longer sufficient, and the circulatory sys-tem diverts the full supply to the lungs. With these alterations made, the newborn's circulatory system closely resembles that of the adult. The biggest difference lies in location and size. An infant's heart sits higher in the chest than an adult's, but soon positions itself lower. Just as the adult heart is about the same size as the adult's fist, an infant's heart is as big as the infant's fist. As the child grows, the heart size maintains about the same proportion.

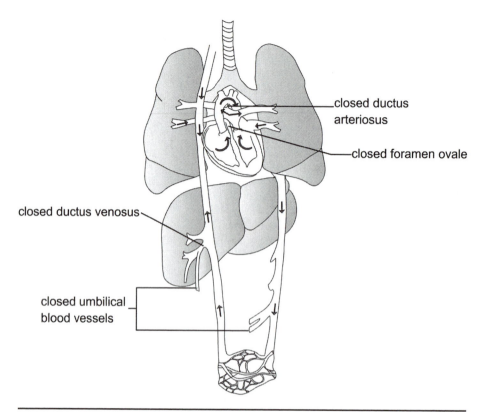

closed ductus
arteriosus

closed foramen ovale

closed ductus venosus

closed umbilical
blood vessels

Figure 6.2. Circulation in a newborn.
All three fetal structures—the ductus arteriosus, foramen ovale, and ductus venosus—close at birth as the infant's circulatory system is separated from the mother's.

IMMUNITY

Antibodies, or immunoglobulins, are one of the body's main lines of defense, conferring an immune response to invading antigens. Immunity is often acquired, which means that as a person faces a new antigen, the body's B cells respond by making antibodies to that specific antigen. The body then "remembers" that antigen, and any future invasion is quickly met with a force of antibodies to fight it off.

For newborns, the world is filled with new antigens. They do, however, have defenses. These include the immunoglobulins IgG and IgA. IgG is transferred from mother to child either during pregnancy or nursing, and IgA is transferred during nursing. The transfer of these immunoglobulins is one of the primary reasons that physicians recommend that mothers breastfeed their infants. Breastfeeding is particularly important immediately after birth when the mother is producing **colostrum**. Colostrum is the fluid that is secreted by the mammary glands right after childbirth and before they begin to produce milk. This fluid is laden with antibodies. As the infant ingests the colostrum, these antibodies pass into its bloodstream and provide resistance to many, if not all, of the diseases to which the mother has already built immunity. These stopgap measures keep the baby healthy in its first months when it would otherwise be bombarded with pathogens in its new environment.

A Picture of the Fetal Heart

Most people are familiar with ultrasounds that allow parents to see their growing fetus inside the mother's womb. While that procedure may provide a glimpse of the fetal heart, medical professionals sometimes require a better view. For that, they often turn to fetal echocardiography.

Like the more standard ultrasound, echocardiography uses high-frequency sound waves to provide an image of the fetus. The procedure displays a real-time, moving image of the fetal heart on a video screen. Usually, an echocardiogram specialist will wait until about halfway into the pregnancy before performing the procedure. At that point, the heart is sufficiently developed to provide a good view of its structure, as well as details about its function.

A physician may prescribe an echocardiography if the fetus exhibits an irregular heartbeat, if either parent has a family history of congenital heart disease, if the mother has diabetes or if she has taken certain drugs during the pregnancy, or if any of a number of other abnormalities is present.

By identifying any problems early on, medical professionals can take pre-birth steps and make post-birth arrangements. For example, they may prescribe medications during the pregnancy to treat arrhythmias. Knowledge of cardiac problems also gives the healthcare team time to make all of the necessary preparations to care for the newborn, and to prepare the parents for any special attention the child may need or potential surgeries that may be required.

7

Blood-demanding Organs and Circulation-related Systems

Blood is almost everywhere in the human body. It flows to all of the tissues, moves in and out of organs, and participates in all sorts of bodily functions. This chapter will introduce some of the organs that put a high demand on blood flow, as well as the lymphatic system that has a close bond with circulation.

DIGESTION

The main arteries feeding the stomach include the celiac, gastric, gastroduodenal, and gastroepiploic arteries (see color insert). The **celiac artery**, or celiac trunk, stems from the abdominal aorta, which is the portion of the aorta that lies in the abdomen (as opposed to the thoracic aorta that runs through the chest cavity). The celiac artery branches into the left gastric, common hepatic, and splenic arteries. For the digestive system, the left gastric artery supplies blood to the stomach and the lower part of the esophagus, which is the feeding tube extending from the pharynx at the back of the mouth to the stomach. The **gastroduodenal artery** separates from the common hepatic artery, which itself continues on to the liver as the hepatic artery. The gastroduodenal artery, the right gastroepiploic artery that branches from it, and the left gastroepiploic artery that arises from the splenic artery all provide blood to the stomach and duodenum (the first part of the small intestine). The right gastric artery arises not directly from the celiac artery like the left gastric artery does, but from the hepatic artery proper. It eventually connects with the left gastric artery and supplies blood to the stomach.

Blood to the colon comes from the left, middle, and right colic arteries, all of which branch from either the inferior or superior mesenteric arteries. Both mesenteric arteries separate from the abdominal aorta. The mesenteric arteries also supply blood to the rest of the intestinal system through the **intestinal**, **ileocolic**, and other arteries.

How does the food a person eats wind up in the blood? The answer lies in tiny outgrowths, called villi (the singular is villus), that line the inside of the wall of the small intestine (see Figure 7.1). These villi are, in turn, coated with even smaller outgrowths, called **microvilli**. By the time digested food reaches the small intestine, it has already been broken down into molecules small enough to cross the capillary walls. Each of the villi has a set of capillaries, thus providing ample opportunity for the uptake of food from inside the small intestine into the blood system. Although the work of the villi might seem inconsequential, they are so numerous throughout the small intestine and are covered by such a large number of microvilli that they actually increase the surface area of the interior small intestine by

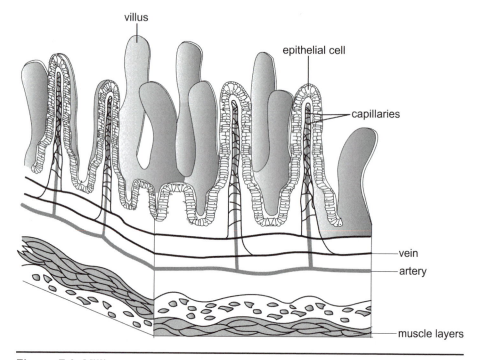

Figure 7.1. Villi.
A multitude of tiny outgrowths, called villi, line the inside of the wall of the small intestine. These villi are, in turn, coated with even smaller outgrowths, known as microvilli (not shown). Each villus has its own set of capillaries, thus providing ample opportunity for the uptake of food from inside the small intestine into the blood system.

about 600 times. This puts the blood into close contact with much of the digested material, or chyme, before it makes the approximately two-hour journey through the small intestine.

The mesenteric veins and other feeder arteries serve as the exit route for blood from the intestines and colon. Gastric and gastroepiploic veins drain the stomach into a number of other veins that ultimately unite—along with the mesenteric veins—at the large portal vein. Thus, nutrient-laden blood leaves the digestive system through the portal vein, but before it begins its return trip to the heart, it heads to the liver.

LIVER AND HEPATIC CIRCULATION

The liver is a large, broad organ (sometimes referred to as a gland) that sits below the lungs and diaphragm, but above the stomach. Measuring about 8–9 inches (20–23 centimeters) long and 6–7 inches (15–18 centimeters) high, this two-lobed, wedge-shaped organ plays an important part in the use and storage of food energy.

About a quarter of the entire cardiac output passes through the liver, but unlike many organs, the liver gets only 25–28 percent of this blood directly from the aorta or its primary branches. That amount, however, is enough to support the organ and its functions. Most of the remaining 72–75 percent of the blood arriving at the liver has already been through the digestive system, where it picked up nutrients from food. This indirect type of blood distribution is termed **in-series blood circulation** because the blood goes from one organ to another in a series. It is also called **portal circulation**.

As soon as the blood arrives in the liver from the portal vein, the liver gets to work removing the food molecules, including fats, the amino acids from proteins, and sugars, and converts them into a carbohydrate called **glycogen** for storage as a reserve energy supply for the body to use later as the need arises. The sugars, called glucose, switch readily to glycogen, but the fats and amino acids must first be converted to glucose before they make the change to glycogen. Because the liver is capable of converting glycogen back into glucose, it serves as a regulator of the **blood-sugar level**.

The liver is also one of the body's lines of defense against drugs, alcohol, poisons, pollutants, and other chemical threats. Just as it filters out sugar, the liver selectively removes these substances via the urine or the bile.

In addition to its responsibilities in metabolism and detoxification, the liver makes other compounds, including cholesterol. For more information about cholesterol and its effects on the cardiovascular system, see Chapter 10.

Blood exits the liver through the hepatic vein that joins the inferior vena cava for the return trip to the heart.

KIDNEYS AND RENAL CIRCULATION

As described previously, the kidneys are the body's blood filters and the site where urine is produced. The two kidneys, shaped appropriately like kidney beans, are each about 4 inches (10 centimeters) long and 2 inches (5 centimeters) wide, and are located in about the middle of the back, with one on either side of the backbone.

Blood is approximately 80 percent water, so the body requires a competent system to maintain a proper water balance. The kidneys provide that service, removing excess water and also filtering out waste products. The water and waste products, including soluble materials like salts, are all eliminated via the formation of urine, which exits each kidney through a long tube, called a **ureter**, that flows into the bladder.

The filtering process begins when blood—about a fifth of the total cardiac output—arrives at each kidney from a separate renal artery (see Figure 7.2). **Interlobar arteries** branch from the renal artery to disperse the blood throughout the kidney and to networks of capillaries, called glomeruli (glomerulus is the singular). The glomeruli come into contact with the kidney's filtering units, called **nephrons**, which are each composed of a twisting epithelial tube. On one end, the epithelial tube eventually empties into the ureter, and on the other end, terminates in a bulb. The bulb, known as a **Bowman's capsule**, surrounds the glomeruli and provides an efficient transfer site for water and waste products to move from the blood to the urinary system. Blood leaves the Bowman's capsule in one vessel, but that vessel quickly branches into a second set of capillaries. This time, the capillaries form a web that weaves around the nephron tubules, which participate in returning much of the water and many of the solutes to the blood through the capillaries. Excess water and solutes continue through the nephron to the ureter. Blood exits each kidney through a renal vein.

In a single day, a single person's kidneys can filter out some 40 gallons (151 liters) of water. Reabsorption returns about 39.6 gallons (150 liters). The difference, 33.8 ounces (1 liter), is about the same as the amount of water ingested by that person in a 24-hour period. Thus, the body neither adds nor loses water in a typical day.

Right and left renal veins drain the kidneys. These veins also transport blood from the adrenal glands that sit atop the kidneys. (See the Urinary System volume in this series for more information on the kidneys.)

SPLEEN

The spleen is a mainly red, heart-sized organ that sits to the left and below the stomach. As noted in the section on the digestive system, one of the

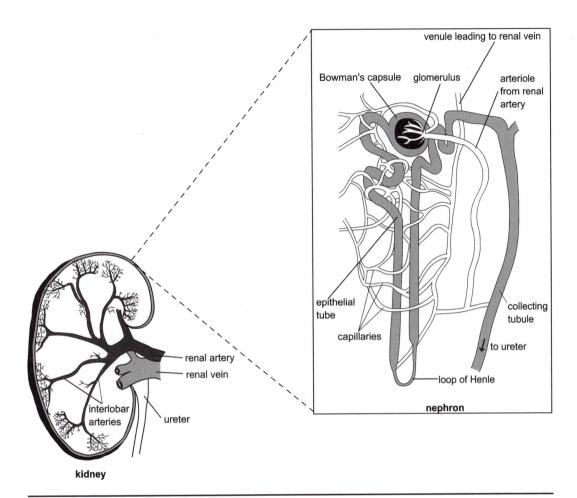

Figure 7.2. Blood flow in the kidney.

A renal artery arrives at each kidney, and then splits into interlobar arteries, then arterioles, and finally to networks of capillaries called glomeruli that are housed in bulbs, or Bowman's capsules. The glomerulus and Bowman's capsule are part of the kidney's overall filtering unit, known as a nephron, which extracts water and waste products from the blood. Much of the water is reabsorbed into the blood through a system of epithelial tubes (also called convoluted tubules). As the water exits, waste products in the tubes, including the U-shaped portion called the loop of Henle, become further concentrated, eventually entering collecting tubules and then the ureter for elimination from the body.

branches from the celiac artery is the splenic artery. This artery supplies blood not only to the stomach, but also to the spleen.

This organ has two primary functions: assisting in the removal of foreign materials and aging red blood cells from circulation, and storing red blood cells and platelets.

To spot and eliminate the threat from invading organisms, the spleen relies on the immune system. The organ contains many branching blood vessels that are enveloped with B lymphocytes and T lymphocytes (B cells and T cells). The T cells have a mission of surveillance, and scan the slowly flowing blood for the foreigners, like bacteria. They report invaders to the B cells. Any B cell that has previously encountered the bacterium rapidly multiplies and then begins producing antibodies to weaken or destroy the invader. The spleen is also armed with many **macrophages**, the white blood cells that ingest and digest bacteria, other foreign organisms, platelets, and old or deformed red blood cells.

In addition, the spleen can swell to hold blood in storage. This ability provides a reserve supply that can be tapped when necessary. Besides the filtration and storage functions, the spleen has another job in the human embryo: It makes red blood cells, a function that shifts to the bone marrow after birth.

CEREBRAL CIRCULATION AND THE BLOOD-BRAIN BARRIER

As discussed previously, the maintenance of blood flow to the brain is one of the circulatory system's highest priorities. This is because the brain is key to so many vital physiological processes. The main arteries to the head include the left common carotid that branches directly off of the aortic arch, and the right common carotid that branches indirectly from the aortic arch by way of a short brachiocephalic (also called innominate) artery (see Figure 7.3). The two carotid arteries traverse almost straight up, branching again at about chin level into internal and external carotids. Other arteries feeding the head include the **vertebral arteries**. These two arteries, one on each side of the neck, arise from the right and left subclavian arteries that divert from the aorta. The vertebral arteries unite at the **basilar artery**, and this artery joins with other cerebral arteries to form what is known as the **circle of Willis**.

Blood supply to the brain comes from numerous major arteries, as well as smaller, branching arteries and arterioles. For example, the internal carotid artery supplies blood to the anterior brain, and one of its branches, called the anterior cerebral artery, feeds the cerebrum. The cerebrum comprises the two large hemispheres of the brain. In some cases, several arteries may supply the same area of the brain. The cerebrum also receives blood from the posterior cerebral artery, which derives from the basilar artery that arises from the vertebral artery. Blood flow to the brain, then, emanates from numerous arteries and arterioles.

Similarly, blood drains from the brain via a large number of venules and veins that empty into several large veins. These include the vertebral vein, and the internal and external jugular veins. The internal jugular is by far

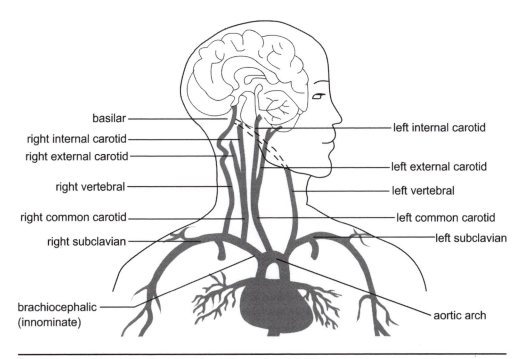

Figure 7.3. Major arteries to the head.

the largest of the three, and runs almost down the middle of the neck. It serves as the primary collector for deoxygenated blood, which it delivers to the subclavian vein and eventually to the superior vena cava for its return to the heart.

Although the brain only accounts for 2 percent of an average person's weight, it demands approximately one-fifth to one-sixth of the cardiac output, or a flow of about 0.74 quarts (700 ml) per minute in an adult. The kidney and liver both require a high percentage of the cardiac output for such maintenance functions as water filtration or waste removal. The brain, on the other hand, needs the large quantity of blood for the oxygen. This heightened demand stems from the extreme rate of oxidative metabolism seen in the nerve cells, or neurons, of the brain. These neurons, collectively called **gray matter**, account for about 40 percent of the brain and use almost 20 percent of the oxygen that a resting person breathes. Just 1 square millimeter of gray matter can have up to 4,000 capillaries! In addition, the gray matter is damaged very quickly if the oxygen supply drops off. Fainting is one of the body's responses to this condition of hypoxia (low oxygen levels). This brings the head closer to the ground, which serves to take advantage of gravity and allow blood to flow more readily to the brain. Continued hypoxia lasting more than a couple of minutes can cause permanent brain damage.

BLOOD-BRAIN BARRIER

The blood-brain barrier is actually a collection of tightly enmeshed cells and other obstacles that effectively serve as a boundary, allowing only oxygen, certain nutrients, and a few other items to pass into brain tissues via the blood (see Figure 7.4). Researchers Paul Ehrlich (1854–1915) and Edwin Goldman saw in the late 1800s to early 1900s that dyes injected into the brain and cerebrospinal system would only color blood there, and dyes injected elsewhere would color all the blood except that in the brain and cerebrospinal system. A barrier existed that selectively prohibited various toxins and other substances from entering the brain. This is now known to protect the brain and its myriad nerve cells from diseases and chemicals that might impair its function. It also, however, prevents many helpful medications from penetrating into the brain, and researchers have been studying how to circumvent the barrier for several decades.

MALE AND FEMALE REPRODUCTION

Besides fetal circulation, which was discussed in Chapter 6, the bloodstream is a large component of the reproductive systems in both males and females. It is a basis for the menstrual cycle in females, and penile erection in males.

Female

Approximately once a month, females of childbearing age but who are not pregnant experience **menstruation**, or the sloughing of the lining of the **uterus**, the organ that holds and nourishes the developing fetus. This lining, also known as the **endometrium**, receives blood from numerous arteries that would help sustain an embryo and fetus following conception. If fertilization doesn't occur, hormones called follicle-stimulating hormone and luteinizing hormone—FSH and LH, respectively—fall to low levels and trigger the endometrium to disintegrate and slough off over a typically three- to seven-day period. About 1.183–1.67 ounces (35–50 milliliters) of blood from the supplying arteries also exits with the lining. A new endometrium begins to form almost immediately, and the cycle repeats unless the woman becomes pregnant or enters **menopause**. After menopause, which commonly occurs from about 50 to 55 years of age, the monthly menstrual cycle ends and the woman is no longer able to conceive.

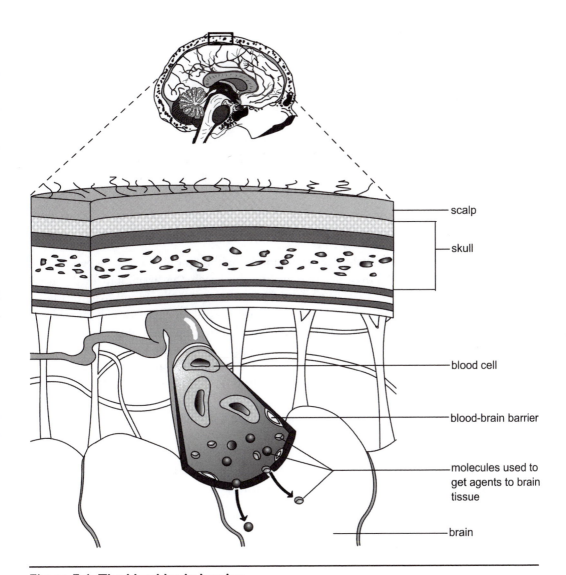

scalp

skull

blood cell

blood-brain barrier

molecules used to get agents to brain tissue

brain

Figure 7.4. The blood-brain barrier.
The blood-brain barrier is a protective boundary that allows only certain gases, nutrients, and other materials to pass from the blood into the brain tissues. In some cases, special molecules help to transport materials across the barrier. Scientists are now learning more about these transportation avenues and other ways to circumvent the barrier to allow beneficial materials, like drugs, to enter the brain. They are also studying why such dangerous entities as bacterial meningitis are able to circumvent the barrier.

Male

The male reproductive organ, or **penis**, requires blood flow in order to become erect for penetration into the female reproductive opening, or **vagina**. Erection occurs because of a combination of nervous and vascular actions.

Sexual arousal stimulates a nervous response that causes the smooth muscle in the arteries and arterioles of the penis to relax and allow blood to flow freely to the two columns of erectile tissue, called the **corpora cavernosa**, on either side of the penis. At the same time, the veins and venules constrict and no longer allow the blood to exit. In other words, blood streams to the erectile tissue, but can't exit. As a result, the tissue becomes engorged with blood, and the penis continues to lift and stiffen to a set point, after which blood inflow ceases. Following the erection, the veins dilate, blood rushes from the corpora cavernosa, and the penis returns to its normal, flaccid condition.

LYMPHATIC SYSTEM

The lymphatic system is sometimes called a "parallel circulatory system" because it has its own network of lymph vessels that transport fluid, called lymph, to the circulatory system. Lymph is much like blood, except that it has no red blood cells, has fewer proteins, and is a bit thinner than blood plasma. Lymph is actually the interstitial fluid that exits the capillaries and enters surrounding cells during the capillaries' exchange function. On average, the lymphatic system returns 11.26 ounces (about a third of a liter) of lymph to the circulatory system every hour. That's enough to fill four 2-liter soda bottles every day.

The lymphatic system collects and transports lymph in a similar fashion to how the circulatory system transports blood. Lymphatic capillaries, which are made of a single endothelial layer, accumulate lymph. The lymph then travels to collecting vessels that are composed of smooth muscle cells. Lymphatic vessels rely on vessel contraction, as well as muscle contraction in the body, to squeeze the lymph and provide the force for the flow. They also have one-way valves to keep the plasma moving in the right direction, just as many blood vessels have valves to maintain a directional flow. The lymph returns to the blood in one of two ways: (1) It may enter through blood capillaries, or (2) it can route through efferent lymphatic vessels to arrive ultimately at the left or right subclavian vein.

Besides lymph vessels, the system has **lymph nodes**. These nodes serve as filters for lymph from the larger vessels, and separate out invading organisms and other foreign material that may have entered the porous lymphatic capillaries. (For a complete description of the Lymphatic System, see the associated volume in this series.)

Into the Brain

Researchers have been struggling for years with the blood-brain barrier. While it is important in protecting the brain from dangerous invading organisms and foreign materials, this barrier also prevents numerous medications that would have therapeutic benefit to the brain from reaching it through the normal delivery route: the bloodstream.

Recently, understanding of the barrier has become precise enough for scientists to devise ways to circumvent the barrier—at least in animal studies. For example, drugs may be attached to and cloaked by molecules that can naturally cross the barrier, or they may be modified so they look enough like other admissible molecules to get past the barrier. Human studies are now underway on methods that temporarily shrink or make gaps between cells in the barrier, effectively creating holes for drugs to use.

For example, researchers at Texas Tech University Health Sciences Center are studying a transport system made of proteins that can shuttle drugs across the barrier. Remarked researcher Dr. David D. Allen of the university, "There are a lot of beneficial drugs that cannot get into the brain because of the barrier. We use these transport proteins, such as the choline transporter, to deliver drugs to the brain." If it leads to practical results, the research could open the doors to new treatments to diseases that affect the brain, including multiple sclerosis, AIDS, and brain tumors.

Scientists are also studying how some dangerous entities, like bacterial meningitis, are able to bypass the blood-brain barrier and enter the brain. According to new findings by a research group at the University of California, San Diego, this form of meningitis is caused by a pathogen known as Group B Streptococcus that may sneak through the barrier by camouflaging itself within a capsule. Said researcher Dr. Victor Nizet, "It appears that the normal GBS with a capsule may represent a form of 'molecular mimicry,' where the bacteria disguises itself to look more like the host and avoid immune recognition." They hope to use their findings to understand how these types of infections occur and develop ways to stop them.

8

The Path toward Understanding: Ancient Philosophers to the Twentieth Century

Just about as far back in history as is recorded, humans have been fascinated by the workings of the human body, including the circulatory system. The Egyptian and Chinese cultures were among the first to place importance on the heart and blood. In at least 2500 BCE, perhaps earlier, papyrus from Egypt indicates that the heart pulsates and is the central distribution point in the body from which a network of vessels disperses. Their description wasn't completely accurate: They also believed that air, water, and semen were transported along those vessels, in addition to blood.

The Chinese came to similar conclusions. Huang Ti (c. 2600 BCE), generally considered the father of Chinese medicine, prepared the *Nei Ch'ing*, a compendium on the human body and various ailments. The *Nei Ch'ing* described blood as a life-giving fluid that is pumped and regulated by the heart and that flows in a continuous circuit through the body via vessels. Other notions were incorrect, however. For example, the Chinese believed that the vessels also transported air, urine, and waste. Nonetheless, the ancient Chinese made considerable strides considering that human dissection was not practiced. Members of the medical community provided care by asking patients about their symptoms and lifestyle, carefully noting details of the patient's body and nuances of the voice, occasionally physically examining the patient and taking the pulse. The latter was much more complicated than a quick reading of the number of beats per minute as is common in doctor offices today. Rather, Chinese medical practitioners of the time not only counted beats, but also recorded time of day, season of year, the practitioner's own pulse for comparison, and subtle differences in individual beats. According to the ancient Chinese—and a ten-volume treatise called the *Muo-Ching* (c. 2500 BCE) on the subject—different parts of the overall pulse could be associated with separate organs, and with numerous diagnoses and prognoses, and would provide clues to the best treatments.

500 BCE TO 200 CE

Human understanding of the circulatory system moved forward from 500 BCE to 200 CE with many of the great philosophers, including Hippocrates, Aristotle, and Galen, exploring human anatomy and physiology and making important contributions to the progress of medicine. Most of the early understanding of the cardiovascular system was tied to what now seems an odd classification scheme. Western philosophers categorized the Earth and its living systems into divisions. Greek philosopher Empedocles of Agrigento in Sicily (c. 493–433 BCE) divided matter into earth, fire, air, and water. These four units were the root of all things, he asserted.

Hippocrates

Hippocrates (c. 460–377 BCE) accepted Empedocles's four roots, as well as the division of the body into its own four categories, called "humors." The humors were phlegm, black bile, yellow bile, and blood. When the humors are out of balance, Hippocrates and others hypothesized, a person becomes sick. Under this scheme, a person with a fever, for example, would be seen to have too much blood in his or her system as compared to the other three humors, and the physician would bleed the patient to bring the four humors back into equilibrium. Although the idea of humors seems bizarre now, it became an underlying tenet of Western medical and scientific thought for 2,000 years.

Some of Hippocrates's other ideas are still with us today in Western medicine. For example, Hippocrates proposed that a person becomes sick for a reason that has nothing to do with magic, quite a radical point of view in a time when people believed that sickness was a punishment for ill deeds or thoughts. Hippocrates also developed the Hippocratic oath, which is still used today as the code of ethics for medical practitioners. He also advocated methodical clinical practices that included patient observation and record keeping. Despite all of his contributions, however, Hippocrates made mistakes. He dismissed the dissections of human corpses as "unpleasant if not cruel" experiments that provided little if any information than that produced by a thorough, external examination of a living patient. His distaste for dissections led to numerous errors in his writings. In his *On the Nature of Man* (c. 400 BCE), for example, he portrays four pairs of large blood vessels in the human body: The first runs from the back of the head down the spine and to the feet, the second leaves the head and travels to the loins, eventually ending in the feet; the third runs from the temples to the lungs, then through various organs, terminating at the anus; and the last pair exits from the front of the head out to the arms and fingers before backtracking to the chest, where each winds its way to the spleen and liver, and finally across the abdomen. He also depicted the heart as containing air in its atria. Still, Hippocrates and particularly his influence on the medical profession have made him one of the most well-known and respected physicians of his time.

Aristotle

About fifty years after Hippocrates made his initial mark on the medical and scientific communities, Aristotle (384–322 BCE) made his own pronouncements about the human body, and the cardiovascular system in particular. A natural scientist rather than a physician, Aristotle nonetheless embraced animal dissection as a tool for inferring the anatomy and physiology

Hippocrates (c. 460–377 BCE), one of the most well-known and respected physicians of his time. © National Library of Medicine.

of the human body. Based on these dissections of "lower" animals, he determined that the heart was the most important organ in the body and that blood vessels originated from the heart rather than from the head or brain, as some other noted philosophers believed. He also became the first to note the existence of the aorta in his writings. He wrote in *On the Parts of Animals* (c. 350 BCE), "[T]he motions of the body commence from the heart, and are brought about by tractions and relaxation." While he made many important discoveries about animal anatomy, he made a number of mistakes, often because he based his understanding of the human body on that of the animals he dissected. For example, he believed that the human heart had three chambers rather than four. He did, however, readily acknowledge that his work was based on animal comparisons and that there was yet much to learn about the human body. Although it is not true in man or other animals, he also asserted that the heart was the only organ in the human body to contain blood, and like many other philosophers, including noted Greek physician Erasistratus (c. 300 BCE), Aristotle thought that the vessels contained and transported air. In fact, the word "arteries" is derived from the term for air ducts. The fact that so many early scientists believed the arteries contained air is not so far-fetched as it now seems. After all, dissections of dead bodies—human or other animal—would not reveal any blood in the arteries. The heart's last beat would have pushed the blood through the arteries and into the veins, which would distend to hold it. The arteries of the corpses, then, would be empty. Based on this finding, it is no surprise that arteries were assumed to carry air.

Herophilus

One of the first Western scientists to declare that arteries carried blood—and only blood—was Herophilus of Chalcedon (c. 300 BCE). Like Aristotle, Herophilus studied in the famed city of Alexandria, Egypt, where most advancements in early Western scientific thought arose. Herophilus obtained much of his knowledge from human cadavers, which he dissected in public. Through this work, he also learned that arteries and veins are different, with the former being noticeably thicker.

Galen

After Herophilus, the understanding of human anatomy and physiology lulled a bit. Human dissections were frowned upon, so new understanding slowed to a trickle ... until Galen. Galen (c. 130–200), also known as Claudius Galenus) was a Greek physician who would come to be considered the father of experimental physiology. Human dissection was still disfavored, but Galen worked around that hurdle by performing many animal dissections. He also served four years as chief physician to the gladiators, which allowed him to carefully examine their often-massive injuries and thus learn from the open wounds how the human body works without having to perform human dissections.

These experiences led him to conclude that experimentation through methods such as dissection was key to learning about human physiology. This was a radical idea in an era when his peers spent the bulk of their time only interpreting the assumedly incontrovertible works of the former great philosophers, rather than conducting their own experiments. This is not to say that Galen lacked respect for earlier scientists. Rather, he held them in high esteem. The difference was that Galen believed medical and scientific knowledge should not end with those great thinkers, but should instead build upon their ideas and findings. Galen wrote about his philosophy, as well as his discoveries about anatomy and physiology, in more than 700 books and treatises, including the influential publication titled *On the Natural Faculties*, and his noteworthy seventeen-volume set, *On the Usefulness of the Parts of the Body*. He wrote both in the second century CE.

Galen took a special interest in the heart and blood, and, like Herophilus, challenged the then-current view that the arteries carried air by asserting that they transported blood. He also determined that the heart was a muscular pump that would continue to beat and move blood

Galen (c. 130–200) performs a surgical procedure on a live pig during a demonstration. Venetian woodcut, 1586. © National Library of Medicine.

even without nervous control, and made spectacularly detailed anatomical observations, including descriptions of veins that adjoin the heart. In addition, he saw a connection between food and blood, hypothesizing that the liver removed waste products from the bloodstream and then transformed food into blood. Although we now know that food doesn't turn into blood, this link eventually spurred scientists, particularly William Harvey (1578–1657) nearly 2,000 years later, to consider a relationship between the circulatory system and metabolism. Other mistakes of Galen included his belief that the blood moved back and forth, rather than in a one-way circuit, and that blood moved from one side of the heart to the other through small pores in the septum.

Although Galen spent his life extolling the virtues of experimentation and advancing current understanding, one of his own works stalled medical progress for nearly 1,800 years. In *On the Usefulness of the Parts of the Body*, Galen employed a strongly religious tone and stated that the creator designed human organs that could perform their roles perfectly. After Galen's death, the Christian church hailed him as the standard authority on medicine, effectively ending any need for further experimentation or medical advancements. This pronouncement, combined with a continued distaste for human dissections, led to hundreds of years of stagnation. In this ironic twist of fate, Galen, the proponent of experimentation, had thus become the latest in a string of "great philosophers," most of whose works would be accepted as the ultimate truth and interpreted by many future generations. A number of his findings about the cardiovascular system, including his descriptions of blood movement from heart to lungs and back, were attributed to animals rather than humans, and thus left for later scientists to rediscover.

1200–1550

Medical thought progressed little from 200 BCE to the mid-1500s with three particularly notable exceptions: Ibn al-Nafis, Mondino de Luzzi, and Leonardo da Vinci. Only one of the three, however, had a wide influence.

Ibn al-Nafis

Ibn al-Nafis (d. 1288), also known as Ibn Nafees, was a prominent Syrian physician with an interest in the cardiovascular system and its possible connection to the lungs. Through his studies, he became the first person known to discover and describe the pulmonary circulation. He also discounted the current thought, which persisted from Galenic times, that blood flowed between the two ventricles via a perforated septum. Instead, he maintained, the ventricles were not directly connected. Rather, blood leaving the right ventricle went to the lungs, and once it traveled through those organs, it flowed back to the heart and into the left ventricle. Ibn al-Nafis fell outside the traditional European medical community, however, so his work remained largely unnoticed to Western philosophers and physicians.

Mondino de Luzzi

Another medical maverick was Mondino de Luzzi (1276–1326), who took it upon himself to conduct the first modern human dissection in 1315, ending the nearly 1,700-year sanction. He led the dissection of an executed woman at the University of Bologna, where he was a teacher. The dissection was apparently performed as a teaching aid for his medical students. Mondino did not actually wield the scalpel, but rather sat separated from the embalmed corpse and read from one of Galen's books as his assistant made the incisions. Mondino purportedly performed some later dissections himself, but often only supervised. With this return of human dissection, Mondino wrote a short, forty-four-page textbook on anatomy and dissection techniques. He penned the book, called *De Omnibus Humani Corporis Interioribus Membris Anathomia*, in 1316, but it remained unpublished until 1487. Notwithstanding the initial delay, a great demand for the book resulted in dozens of editions. Mondino became known as the "restorer of anatomy."

Mondino no doubt saw differences between Galen's descriptions of the human anatomy and what he observed through his own dissections, but he never noted those differences, no matter how obvious they were. For example, he accepted the dogma that numerous channels connected the stomach to the gall bladder and spleen, even though his dissections revealed no such connections. Galen's work had become so ingrained as the ultimate truth that even though Mondino was able to accept Galen's ideas about experimentation being critical to medical understanding, he was unable to follow the master's other primary precept of building upon the work of others, and he rationalized away what he saw before his own eyes.

Although he had his shortcomings, Mondino paved the way for future human dissections. These were performed mainly on executed criminals, although in the absence of enough subjects, medical students were known to sneak out to a nearby cemetery and dig up a corpse.

Leonardo da Vinci

Leonardo da Vinci (1452–1519) is known as an inventor and artist, but he was also an accomplished anatomist who performed many dissections of animals and later humans. His skills as an artist allowed him to make amazingly detailed sketches of various organs, including the heart. Through his dissections, he provided many new insights: The heart doesn't warm the blood as was previously held; it has four chambers rather than three; it has valves that swing open and shut to aid in blood flow; and the contractions of the left ventricle are associated with the wrist pulse. Unfortunately, Leonardo never shared his anatomical studies with the wider medical community. Only generations later were his artistic renderings found and his genius appreciated.

THE MEDICAL RENAISSANCE

The Renaissance was a period of reawakening. For the arts in particular, the transition from the Middle Ages to a modern era began in the 1400s, mostly in Italy, then spread throughout Western Europe. Medical thought lagged behind a bit, and really only began to enjoy a rebirth in the 1550s that culminated in the 1600s. Some of the most important advances in medical history—including the first major publicized discoveries about the cardiovascular system in nearly two millennia—occurred during this period.

Michael Servetus and Matteo Realdo Colombo

Although Ibn al-Nafis three centuries earlier had described the pulmonary circulatory system and discounted any perforations in the septum that linked the right and left ventricles, this knowledge still hadn't made its way into European thought. Apparently unaware of the earlier work on pulmonary circulation, Michael Servetus (1511?–1553), and Matteo Realdo Colombo (1516?–1559) made similar pronouncements. Both announced that blood flows from the right heart to the lungs, then to the left heart and to the body. Servetus rejected the traditional assertion of septum perforations. He also surmised that the pulmonary artery from the right heart to the lungs carried not just some but all of the blood from the body to the lungs, where it was somehow changed, then returned by veins to the heart for distribution back to the body. He likely shared his findings with his peers, but he sidestepped wider acknowledgment by publishing them only in a manuscript titled *Christianismi Restitutio*, or *Restitution of Christianity* (1533) that dealt mainly with his nonconformist religious views. Within a year of that book's dissemination, Servetus was burned at the stake for heresy.

Through dissections of dead and probably living animals (called vivisection), Colombo also noted valves at the heart ventricles that allowed a strictly unidirectional blood flow, and he pointed out for the first time that the heart contracted and relaxed in a phased cycle.

Andreas Vesalius

Andreas Vesalius (1514–1564) is the man who finally released the stranglehold that Galen's ideas on anatomy and physiology had had on the medical community for some 1,800 years. He was born in Belgium to a long line of physicians noted for their service to royalty. In keeping with family tradition, Vesalius studied medicine, but became particularly interested in learning more about anatomy, and performed many dissections of rodents, insectivores, cats, and dogs.

He entered the University of Paris in 1533 and refined his dissection techniques under Jacobus Sylvius (1478–1555). Vesalius initially praised Sylvius and his traditional Galenic view, but later found him to be immovable in his acceptance of Galen's anatomy even when presented with irrefutable evidence that many of the descriptions provided by Galen of human organs and tissues were incorrect. Said Vesalius in his 1546 *Letter on the China Root,* "It happened one day that we showed him the valves of the orifice of the pulmonary artery and of the aorta, although he had informed us the day before that he could not find them. . . . [Since he] read nothing else anatomical except the books *On the Movement of Muscles* in which he everywhere agreed with Galen, it is not astonishing that I write that I have studied without the aid of a teacher." Vesalius also assisted in dissections supervised by another teacher, Johannes Guenther (1487–1574), who was similarly an ardent believer in Galenic anatomy. Vesalius roundly criticized him as well for simply quoting from ancient books rather than actually performing dissections and learning from them. Nonetheless, Vesalius became a skilled dissectionist of animals and of human convicts, whose bodies he obtained following their executions.

From the University of Paris, Vesalius returned to Louvain, where he had studied in the late 1520s, and then left for Italy's noted University of Padua, where he earned his doctor of medicine in 1537, and became a professor of surgery and anatomy, apparently the first such position at any university. His teaching method included lectures simultaneously accompanied by dissections on human bodies. He became a prolific writer/illustrator, producing *Tabulae Anatomicae Sex,* or *Six Anatomical Tables* (1538), a series of educational, wood-cut plates of the human anatomy. His work became known throughout Western Europe, Vesalius's reputation as a scholar soared, and he soon became emboldened to announce discrepancies in Galen's descriptions of the human body. In 1540, he went to Bologna with the skeleton of an ape and a human, and plainly showed that dozens of Galen's descriptions fit the ape's body, but not the human's. Clearly, he reasoned, Galen's human anatomy was inferred from the dissections of an ape. This public demonstration essentially marked the beginning of the end for the blind acceptance of Galen's anatomy and heralded a period of experiment-based work, something that, ironically, Galen himself had advocated eighteen centuries earlier.

Armed with the success of his public demonstration, Vesalius began to write his most famous work, *De Humani Corporis Fabrica* (often shortened to *De Fabrica*), which was published in 1543. This seven-volume set, including fifteen chapters on the circulatory system, and a second edition in 1555 spelled out the "new" human anatomy, fervently pointing out errors in the traditional point of view both in words and in engravings. One of the revelations that was perhaps most shocking to his contemporaries was his evidence, gained through dissection, that the septum of the heart completely divides the right and left ventricle, and has no connecting perforations as Galen hypothesized. *De Fabrica* was widely acclaimed—and plagiarized—throughout Europe, but the approval was by no means unanimous. Traditionalists reviled Vesalius, ardently and repeatedly criticizing him and his views. Just a few months after he published the first edition of *De Fabrica*, Vesalius became so incensed with the attacks that he resigned his position at the university and, despite the pleadings of friends who were with him at the time, destroyed his considerable unpublished papers. *De Fabrica*, however, stood the test of time. The publication became the centerpiece of Vesalius's career, and for his contributions, he is now known as the "father of anatomy."

Girolamo Fabricius

The work of Vesalius also lived on through his students. Gabriele Fallopio, or Gabriello Fallopius (1523–1562), studied under Vesalius and drew fame for discovering the Fallopian tubes as well as for teaching Girolamo Fabricius, or Fabricius ab Aquapendente. Fabricius (1537–1619) became an anatomist at Padua and concentrated on the circulatory system. He found and described "little doors," the valves in veins. Although he was unsure of their purpose, he felt the finding important enough to warrant the 1603 publication of *On the Valves in Veins*. His most significant contribution, however, might be the foundation of solid research inquiry that he instilled in his own students, including William Harvey (discussed below).

Andrea Cesalpino

Before William Harvey would write his revolutionary treatise on the circulatory system, Andrea Cesalpino (1519–1603) presented the first known arguments in Western medicine that the blood moved through the body in a closed, continuous pathway. He was also the first person to suggest that very fine vessels, now known as capillaries, connected arteries to veins. Despite his insights, he didn't comprehend the full picture of blood circulation as Harvey would.

William Harvey

A British physician and anatomist, William Harvey in 1628 revolutionized the understanding of the human body, and especially the circulatory system, through his famed *Exercitatio Anatomica de Motu Cordis et Sanguinis in Animalibus* (*Anatomical Treatise on the Movement of the Heart and Blood in Animals*). The treatise, typically shortened to *De Motu Cordis*, provided the first accurate and widely accepted picture of the flow of blood through the heart, veins, and arteries.

Born in 1578 near Dover, Harvey earned an undergraduate degree from Caius College (Cambridge) before continuing his education at Padua, where he hoped to study with the masters of anatomy. He indeed became an assistant to Fabricius and no doubt was immersed in investigations of the circulatory system. With his diploma in hand, Harvey returned to London, became a fellow of the Royal College of Physicians, and then became a physician to King Charles I. His true calling, however, was research.

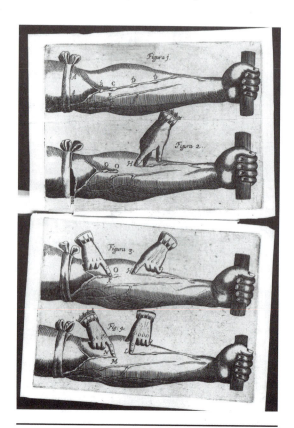

William Harvey (1578–1657) used this and other illustrations in his famous treatise *De Motu Cordis* to explain how blood flows from the heart in a continuous one-way cycle. Courtesy of the Library of Congress.

Harvey conducted many dissections to learn about the cardiovascular system, but found that vivisections provided insights that dissections of dead bodies could not. He performed vivisections on all manner of animals, often cutting into the chests of live pigs during lectures at the Royal College to demonstrate that blood traveled in one direction. By severing veins and arteries, he could show that the blood ran from the right ventricle to the lungs, and then to the left ventricle to the aorta. Through observations of the slow-moving hearts of ectothermic (cold-blooded) animals and of dying endothermic (warm-blooded) animals, he learned that the atria contract slightly before the ventricles. Some of his most revealing experiments, however, involved cutting the blood vessels of living animals, including humans. First, he would tie off an artery, wait a moment, then slice the vessel on both sides of the constriction to see which side contained the blood. Blood on the side closest to the heart but not on the other meant that arterial blood arrived from the heart. He did the same thing with a vein, and showed that a constricted vein held blood on the side away from the heart. This simple but elegant experiment provided the first proof that blood traveled out from the heart to the arteries, but from the veins to the heart. In addition, his dissections clearly substantiated other researchers' observations that the septum between the left and right ventricles was whole, and not perforated.

Harvey used demonstrations to show that the pulse was caused by heart contractions, and that each contraction pumped blood. Through his observations of a beating heart, he wrote in *De Motu Cordis*, "These things, therefore, happen together or at the same instant: the tension of the heart, the pulse of its apex, which is felt externally by its striking against the chest, the thickening of its parietes, and the forcible expulsion of the blood it contains by the constriction of its ventricles." This statement, like many others in the manuscript, challenged current beliefs. In particular, the current view held that the ventricles are filled with blood when the heartbeat is felt. His inspections indicated the opposite.

He was also the first to estimate the vast amount of blood pumped by the heart in a single day and to conclude that the body must use the same blood over and over again. In other words, he said, the blood must move in a closed loop—a circulatory system—that is driven by the heart, which serves as a muscular pump. He likely arrived at the bulk of his many conclusions by 1615, but spent the following years convincing his contemporaries at the college of the truth of his findings before making the claims in *De Motu Cordis* thirteen years later.

In that manuscript, Harvey used the first seven chapters to provide anatomical and physiological details of the heart and blood vessels. Many of these details were based on recent findings made by other anatomists, although he left out most attributions. To these, Harvey added his own enlightening discoveries. While his work was heralded by his peers in London, many traditionalists still clung to the Galenic view of the cardiovascular system. Some claimed that some of the primary findings in his dissections were mere artifacts. For instance, detractors asserted that his dissections showed not that septum perforations are absent, but only that they closed up postmortem. Perhaps the most damaging criticism, however, centered on a gap in Harvey's circulation that he was unable to bridge. Like Cesalpino, Harvey surmised that small vessels linked the arteries to the veins and completed the blood's closed circuit, but he could provide no direct proof. The

Richard Lower (1631–1691), whose observations provided insights into the connection between respiration and circulation. © National Library of Medicine.

development of microscopes was on the horizon, but not soon enough for Harvey. He died four years before one of the earliest microscopists would view the capillary network.

Richard Lower and John Mayow

While Harvey noted the color differences in arterial and venous blood, it was Richard Lower (1631–1691) and John Mayow (1640–1679) who saw its significance. Lower demonstrated that air in the lungs changed the dark, purplish blood in the veins to a bright red fluid in the arteries. Mayow took it a step further and suggested that some type of exchange occurred in the lungs, with the air trading what he called "nitro-aerial spirits" for some unidentified "vapors" in the blood. The hypothesis was the first step toward an understanding of respiration and oxygen–carbon dioxide exchange.

Besides his rudimentary work on respiration, Lower became in 1666 the first person known to perform blood transfusions. In his experiments, he drained the blood from a dog until it became unconscious, then used a long syringe made from the quill of a feather to link its jugular vein to the artery in the neck of another dog. The introduced blood brought the bled dog back to consciousness.

Jean-Baptiste Denis (Denys)

At about the time of Lower's success, French physician and philosopher Jean-Baptiste Denis (1643–1704) was doing similar dog-to-dog transfusions. In 1667, he made the leap to humans and at the Académie des Sciences in France conducted what is believed to be the first blood transfusion in a person. His patient was an ailing 15 or 16 year old who had already been bled several times. Physicians of the day believed they could cure many maladies by removing "bad" blood. Denis prescribed a transfusion, but instead of obtaining blood from another human, he ran a tube from a vein in the boy's arm to the carotid artery of a lamb. The boy survived. Lower began his own trials with transfusions, and attempted a sheep-to-human transfusion, which the patient also survived. Denis continued with the additional lamb-to-human transfusions until the end of the year, when a patient named Antoine Mauroy died. Mauroy's widow apparently made public accusations about Denis's character and/or abilities, Denis sued her for slander, and the French Parliament settled the matter by prohibiting further human transfusions.

Transfusion experiments continued in other countries, however, and by the early nineteenth century, human-to-human transfusions began.

THE ADVENT OF MICROSCOPES

The advent of microscopes in the early seventeenth century gave scientists and physicians a novel view of the human body and its systems, not only fortifying previous assumptions but also opening the doors to a new era of discovery.

Jan Swammerdam

In the beginning, microscopes were crude instruments with poor resolution, but they were still good enough to yield major findings. One of the earliest observations came in 1658, when young Jan Swammerdam (1637–1680) was peering at a blood sample through his microscope. The 21-year-old Dutchman became the first person to view and describe red blood cells. More than two centuries would pass before white blood cells were identified.

Marcello Malpighi

The next major finding regarding the circulatory system came just three years later in 1661. Marcello Malpighi (1628–1694), an anatomist at the University of Bologna, used his primitive microscope to view lung tissue and, within it, a network of tiny, heretofore invisible vessels that stretched from the smallest arteries to the smallest veins. He had found capillaries, the missing piece of the puzzle in Harvey's blood circulation. Malpighi's observation put an end to much of the criticism that Harvey had endured during his lifetime. Even previously skeptical scientists and physicians now had clear evidence that the veins and arteries were connected, and that the blood did, in fact, travel in a closed loop.

Anton van Leeuwenhoek

The name Anton van Leeuwenhoek (1632–1723) is tied to anatomy, but he started out as a dry-goods merchant who had an interest in magnification. He began by making magnifying lenses, which he ground to be more and more precise and powerful until he was able to construct working microscopes. In fact, it was through one of Leeuwenhoek's scopes that Malpighi saw the first capillaries. Despite this tie between the two men, Leeuwenhoek apparently was oblivious to Malpighi's discovery. Also seemingly unaware of Swammerdam's groundbreaking view of the red blood cells, Leeuwenhoek used his apparati to see blood close up as it coursed just beneath the skin of a frog's limb. He was also able to see and provide detailed descriptions of red blood cells, and actually viewed them *inside* capillaries. A highly respected Dutch anatomist named Regnier de Graaf (1641–1673) became aware of Leeuwenhoek's microscopes and reported this new technology to the Royal Society of London. Intrigued, the society invited Leeuwenhoek to submit his findings for publication in its *Philosophical Transactions.*

In his sixty-fifth missive to the Royal Society in the late 1700s, Leeuwenhoek wrote, "I was greatly pleased to see very distinctly the circulation of the blood (in the tadpole), which was driven on from the parts that were nearest to the body to those on the outside, thus causing an uninterrupted, very rapid circulation." He watched blood move in cycles: first swiftly, then slowly. He wrote, "From this, I concluded that as often as these sudden impulses occurred, the blood was driven from the heart." Of the capillary movement he observed in frogs, he noted, "If now we see clearly with our eyes that the passing of the blood from the arteries into the veins, in the tadpoles, only takes place in such blood-vessels as are so thin that only one corpuscle can be driven through at one time, we may conclude that the same thing takes place in the same way in our bodies as well as in that of all animals."

Although their magnification power of 270 times may seen minimal by today's standards, Leeuwenhoek's microscopes remained state-of-the-art instruments for more than 100 years.

1700–1900

The 1700s and 1800s can be seen as a continuation of medical enlightenment, but also the period when physicians became aggressive in putting the basic science to work in the clinical setting. Many of the inventions now seem odd, clumsy, even dangerous, but they nonetheless marked the beginning of many advances in diagnosis and treatment.

Sir John Foyer and the Pulse Watch

Although Western medicine was well aware of the pulse and its connection to disease, physicians still had no standard timing device for quickly and accurately measuring the pulse rate. That changed in 1707 when Sir John Foyer (1649–1734) of England created a "pulse watch." Designed to tick off exactly one minute, a physician could count the number of pulsations while the watch ran. Although the watch didn't catch on, it represented one of the first efforts to mechanize and standardize medical diagnostics.

Stephen Hales

Another clinical leap came when English physiologist Stephen Hales (1677–1761) made attempts to measure blood pressure. In what now seem crude experiments, he punctured the artery of a horse, then slid inside a glass tube. The tube contained mercury, which rose and fell with every heartbeat. Using measurements carefully etched on the tube, Hales recorded changes in the mercury levels, and thus the blood pressure. He announced the technique in 1733, but without being able to develop a system that didn't involve slicing into an artery, his invention was never incorporated as a diagnostic tool for humans. Nonetheless, his work brought attention to the connection between blood pressure and health.

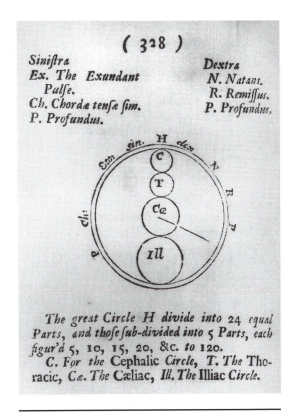

Sir John Foyer (1649–1734) designed the first device to quickly and accurately measure the pulse rate. This diagram shows his device, a physician's pulse watch. © National Library of Medicine.

Luigi Galvani

By the end of the century, a physician and anatomist from Italy was beginning to see a connection between the heart and electrical activity. This was Luigi Galvani (1737–1798). Galvani's name is often associated with his friend Alessandro Volta (1745–1827), a physics teacher, who engaged with Galvani in a lively debate over "animal electricity." At the University of Bologna, Galvani had begun experimenting on frogs by jolting them with electricity and recording their responses. He found that electricity applied directly to the muscles of the frog's leg or even to nerves leading to those muscles would make the leg twitch. While Galvani's work didn't involve the heart, which is also a muscle, it did form the foundation for future understanding of the electrical activity of that organ.

William Hewson

In 1771, British physiologist William Hewson (1739–1774) published his masterwork *Experimental Enquiry into the Properties of the Blood*. The manuscript was the culmination of his years of research on the mechanisms of blood coagulation. Through his experiments, he found that blood would remain a liquid much longer if it was left in a segment of a vein than if it was poured into a bowl. Something was causing the drawn blood to coagulate. Hewson spent countless hours in his lab trying to find the mechanism behind coagulation, and was rewarded when he discovered fibrinogen, a plasma protein that he called "coagulable lymph." By deactivating fibrinogen, particularly through the use of neutral salts, it would not clot. His research began an important foundation for clinical transfusions, which had previously run into problems with blood clotting in the tube and blocking its transferal from patient to patient.

Two years after the publication of *Experimental Enquiry*, Hewson provided the first evidence that red blood cells have a surrounding membrane. He died the following year of blood poisoning sustained while performing a dissection.

René-Théophile-Hyacinthe Laënnec

A stethoscope was the next major new cardiovascular invention. Created in 1816 by French physician René-Théophile-Hyacinthe Laënnec (1781–1826), the device was a simple hollow tube. The invention came about when Laënnec needed to check the chest sounds of a female patient. He felt it too forward to follow the customary practice of putting his ear to her chest, so he rolled up a paper tube and listened through the makeshift device. Finding the chest sounds to be not only audible but spectacularly crisp, he proceeded to refine a wooden cylinder that would be even more sensitive. In his manuscript, *De l'auscultation mediate*, published in 1819, he provided details of the auditory signals for many cardiac and pulmonary diseases.

Laënnec's stethoscope became a very popular diagnostic tool that was used in doctor offices until the next-generation stethoscope appeared in the 1850s. The replacement had earpieces to improve audibility, and used rubber instead of wood for the tube.

John Henry Leacock

Following the 1600s and Jean-Baptiste Denis's ill-fated lamb-to-human transfusion and the resulting French ban on the procedure in humans, the use of clinical procedure in humans stagnated. John Leacock (1793?–1828) of Barbados brought renewed interest to the possibilities when he deduced through careful investigation in dogs, cats, and sheep that blood transfusions should only be made between like species. He suggested that human-to-human transfusions should be explored, although he apparently never performed the procedure himself.

James Blundell

English physiologist James Blundell (1790–1878) took up the gauntlet thrown by Leacock and by 1818 had begun performing the first human-to-human transfusions. He used a syringe-and-pump apparatus to deliver a supply of already-collected human blood to new mothers who had experienced severe bleeding in childbirth. He tried the technique on more than a dozen women, and about two-thirds survived.

Carlo Matteucci and Emil DuBois-Reymond

About three decades after Luigi Galvani noted that electrical activity appeared to play a role in muscle contraction, an Italian scientist and then a German physiologist verified the relationship, even in cardiac muscle. Carlo Matteucci (1811–1868) performed a series of trials beginning in 1830 that showed that electricity is involved in muscle contraction. In 1831, Emil DuBois-Reymond (1818–1896) of Germany was actually able to measure the difference in electrical current between relaxed and contracted muscle, and called that difference an "action potential."

Albert von Kölliker and Heinrich Müller

Albert von Kölliker (1817–1905) and Heinrich Müller (1820–1864) took the findings of Matteucci and DuBois-Reymond further, illustrating through their experiments of frogs that each heart contraction is accompanied by an electrical current, and measuring the action potential. They pursued the line of thought and were able to discern two current changes, the larger of the two slightly preceding the ventricular contraction. With this result, they hypothesized that electrical variation is the trigger for heart contraction.

Kölliker also conducted his own research on blood vessels and found that the muscle tissue comprised fibers quite different from skeletal muscle. These fibers, actually long cells, had no striations and contained nuclei.

Hermann Friedrich Stannius

The pacemaker was identified in 1852 when Hermann Friedrich Stannius (1808–1883), also experimenting on a frog, isolated the sinus venosus by tying tiny knots between it and the rest of the heart. He also noted that the human heart's four chambers would spontaneously resume contracting, indicating the presence of a backup system in case of pacemaker failure.

Sir William Osler

Following William Hewson's discovery nearly a century earlier of the "co-agulable lymph" we now know as fibrinogen, Sir William Osler (1849–1919) scrutinized bone marrow and found tiny bits of cells that also play a role in blood clotting. These fragments, which came to be called platelets, have a key role in the quick-acting series of chemical reactions that result in co-agulation.

Élie Metchnikoff and Hans Buchner

While red blood cells were well known, even being observed through a microscope and copiously described by numerous scientists, white blood cells weren't even conjectured until 1892. In that year, Russian biologist Élie Metchnikoff (1845–1916) proposed that, in addition to red blood cells, the blood contained other amoeboid blood cells (the white blood cells) that attack, surround, and destroy germs. Metchnikoff would go on to receive the 1908 Nobel Prize for his work.

Just a year earlier, Hans Buchner (1850–1902) asserted that the blood contained proteins that would fight germs. Buchner and Metchinikoff argued over cells versus proteins as the basis for the body's antibacterial defense, but both eventually were proved to be correct.

George Oliver and Edward Albert Sharpey-Schäfer

As scientists were beginning to gain an understanding of electricity's effects on the heart, two other scientists announced that something else could trigger muscle contraction in blood vessels. In 1894, George Oliver (1841–1915) and Edward Albert Sharpey-Schäfer (1850–1935) reported that material obtained from the adrenal gland could cause vessel contraction. Their conclusion stemmed from animal experiments and observations of a spike in blood pressure following exposure to the adrenal extract. The responsible compound was later called a hormone.

Ernest Henry Starling and William Maddock Bayliss

The Frank-Starling law of the heart, Starling's equilibrium, and Starling's forces all have one thing in common: Ernest Henry Starling (1866–1927). The work of this British physiologist helped explain how the circulatory system worked, including the previously unknown mechanism of capillary exchange.

Starling prepared for his illustrious career at Guy's Hospital Medical School in London, where he earned his degree in 1889. He then joined a laboratory at University College in London and met the slightly younger William Maddock Bayliss (1860–1924). The two bonded quickly and soon became interested in a new device, termed a "capillary electrometer," that could discern the electrical activity of the heart. Their work built on the recent recordings of electrical activity, which were published by scientists John Burdon Sanderson (1828–1906), Frederick Page, and Augustus D. Waller (1856–1922). Bayliss and Starling tinkered with the device and the placement of its terminals, and were able to demonstrate that each heartbeat has three separate phases, later called P, QRS, and T, and that the ventricles contract approximately 0.13 seconds after the atria.

Starling left University College in 1892, spending time in other labs, including that of Élie Metchnikoff. Starling then initiated a vigorous research project to resolve the mechanisms of capillary exchange. By 1896, he had determined that an equilibrium exists at the capillary level. He suggested that solutions of proteins, called colloidal solutions, in the capillaries regulate interior and exterior salt solutions via osmotic pressure. According to his hypothesis, the proteins in the capillaries are too large to exit the semipermeable capillaries, but dissolved salts can move across them readily. Therefore, when salts in the interstitial space (outside the capillaries) become too concentrated, they pass through the capillary membranes and into the capillaries, bringing tissue fluids with them. If, on the other hand, salts become more concentrated inside the capillaries, they can flow to the interstitial space, taking excess water with them. His hypothesis is now known as Starling's equilibrium, and the osmosis-driven mechanism commonly goes by the name Starling's forces. The overall hydrostatic pressure also plays a role in this equilibrium, because this mechanical force, which can result from elevated venous pressure due to congestive heart failure, also pushes liquid out of capillaries.

Starling's Law of the Heart resulted from his investigations of ventricular contraction. This law explains that the length of myocardial fibers is proportional to the strength of ventricular contraction. Therefore, if the ventricle is distended by an increase in pressure and thus volume, the muscle fibers elongate, and the following ventricle contraction will be more forceful.

In 1899 Starling accepted a faculty position at University College and rejoined Bayliss, who had just returned to the institution after an appointment at Oxford. The two began working together again and identified in 1902 a digestive substance (secretin) that travels through the bloodstream to trigger the secretion of digestive juices into the intestine. Starling later dubbed the substance a hormone. This finding was an important impetus behind the development of the endocrinology field and the comprehension of these chemical messengers.

Willem Einthoven

In 1895 a Dutch physician, Willem Einthoven (1860–1927), further refined the heartbeat's three deflections, which had been identified by Starling and

Willem Einthoven (1860–1927) in his laboratory with his original string galvanometer, used to record the heart's electrical activity. © National Library of Medicine.

Bayliss, and named the now-five deflections as P, Q, R, S, and T. In 1923, he accepted the Nobel Prize for work in developing the modern electrocardiograph (originally developed under the name "string galvanometer") to record the heart's electrical activity. By this time, the instrument had become available in doctor offices and hospitals as a diagnostic tool.

Etienne Jules Marey, Carl Friedrich Wilhelm Ludwig, and Scipione Riva-Rocci

Following Stephen Hales's announcement in 1733 of a method of measuring blood pressure (sliding a mercury-filled tube into an artery), other researchers began devising next-stage instruments. Among these inventors were Etienne Jules Marey (1830–1904), Carl Friedrich Wilhelm Ludwig (1816–1895), and Scipione Riva-Rocci (1863–1937). Marey invented a device called a sphygmograph to record a patient's circulation, and Ludwig created a kymograph to record blood-pressure changes. Riva-Rocci holds his place in history as the Italian doctor who developed the modern blood-pressure cuff, called a **sphygmomanometer**. The device is still in use in doctor offices today.

History of Hemophilia

The symptoms of hemophilia and the fact that it is more prevalent within certain families have been recognized for at least 2,000 years. This genetic disorder causes uncontrolled bleeding, leading to often-severe anemia, and sometimes death from minor injuries that would warrant little more than a bandage in a normal individual.

It wasn't until 1828, however, that the disease got the name hemophilia. This came about twenty-five years after physician John Otto of Philadelphia described the disorder, including its symptoms in males and its hereditary nature.

Hemophilia became better known through its connection to Queen Victoria of England in the nineteenth century. The queen carried the gene for the disease and passed it to a son, who eventually died at age 31 from a brain hemorrhage. Before his death, the son, Leopold, had a daughter who was a carrier, and a son who also died of a brain hemorrhage. Two of Queen Victoria's daughters were also carriers. One of the daughters married into the Russian royal family and had a son with hemophilia.

In 1937, physicians A. J. Patek and F.H.L. Taylor at Harvard University studied hemophilia and found a substance in normal blood that was lacking in the blood of a person with hemophilia. When this substance, which they called "anti-hemophilic globulin," was added to the blood of hemophiliacs, it clotted. By the 1950s, scientists had learned that hemophilia came in two forms, called hemophilia A and B, that resulted in deficiencies of one of two proteins. Hemophilia A resulted from a lack of the protein called Factor VIII, and patients with hemophilia B had no Factor IX.

By the late 1960s and early 1970s, patients began receiving concentrates of the factors to treat the disorder. Unfortunately, these concentrates weren't screened for HIV (the AIDS virus) or hepatitis C, so infections with these and other viruses became common in hemophiliacs. Blood is now screened for these contaminants, vastly improving the outlook for people with hemophilia.

From Flower to Pharmacy

Many of the drugs at the pharmacy have botanical origins, including the well-known heart drug called digitalis.

The path from garden plant to clinical treatment moved forward in the 1700s when physician William Withering (1741–1799) began investigating the properties of leaves from the foxglove plant, *Digitalis purpurea*. Tea made from the leaves was a folk remedy for edema, which is the accumulation of fluids in body tissues. Once scientists linked edema with heart disease, Withering began studying the leaves' effects on the heart. In 1785, he published his work showing that the leaves lowered the heart rate. It is now known that **digitalis**, and specifically active chemicals called cardiac glycosides, also boost the heart's contraction so that blood flow is maintained despite a slower heartbeat. Withering's findings quickly propelled digitalis to the top of the list of heart medications.

Since Withering's discovery, other plants have been found to contain cardiac glycosides. An example is the common milkweed (*Asclepias syriaca*). Botanists and entomologists recognize this plant for its relationship with the larvae of monarch butterflies (*Daneus plexippus*). Caterpillars eat milkweed leaves exclusively and absorb the glycosides, which are toxic to birds and other animals that prey upon the caterpillars and adult butterflies.

Tumors and Blood Supply

When a cancerous tumor begins to form and throughout its growth, it prompts the growth of a web of blood vessels to bring it oxygen and nutrients. If some means were developed to cut off that supply, the tumor would starve to death.

The development of drugs to do just that has become a very active area of research. In fact, proof of the importance of blood supply to tumors came in the 1700s when famed physiologist William Harvey cut off the blood leading to a tumor and eradicated it.

The formation of blood vessels around a tumor is called **angiogenesis**. These vessels not only nourish the tumor but also provide an avenue for cancerous cells to leave the tumor site and spread to other areas of the body. Already, researchers have found several methods that halt blood supply in laboratory models, and are making advancements in new anti-angiogenesis drugs. Some target the pathway involved in making blood vessels, zeroing in on and disrupting critical steps.

9

The Path toward Understanding: The Twentieth Century

Understanding the circulatory system had progressed greatly from the 1500s to the 1900s, but much was still left to discover. The clinical front saw particular advances in the twentieth century, and researchers today continue to improve treatment options, to develop diagnostic tools, and to reveal the still-hidden secrets of the cardiovascular system.

EARLY TWENTIETH CENTURY

Paul Ehrlich

A medical doctor who became committed to research, Paul Ehrlich (1854–1915) quickly rose in the German scientific community, holding positions at the Institute for Infectious Diseases in Berlin (the Robert Koch Institute) and later at the Institute for Serum Research and Serum Testing. Through his work with antitoxins, he would finally end the debate that stemmed from Élie Metchnikoff's and Hans Buchner's ideas about the basis for the body's germ-fighting defenses. Metchnikoff asserted that white blood cells attack, surround, and destroy germs, but Buchner's work indicated that proteins were the body's protectors. In 1903, Ehrlich demonstrated that antitoxins were chemical substances that link to invading bacteria, essentially dismantling their resistance so that the white blood cells can eliminate them via phagocytosis. Five years later, he and Metchnikoff shared the Nobel Prize for their elucidation of the immune system.

Karl Landsteiner

Karl Landsteiner (1868–1943) helped kick off the new century with the important discovery of three human blood groups: A, B, and O. Born in Austria but a naturalized U.S. citizen, Landsteiner systematically mixed different samples of blood and sera, and carefully recorded which combinations caused the red blood cells to clump, or agglutinate. He determined that the blood comes in three types: A, B, and C (later called type O). The sera from type A blood causes type B to clump when they are mixed, and type B sera has the same effect on type A. Neither, however, had any agglutinizing effect on type O. Armed with this data, he deduced that the types can be distinguished by the presence or absence of two antigens and their associated antibodies. Landsteiner published his findings in 1901. In the following year, Landsteiner's fellow researchers Alfred von Decastello and Adriano Sturli (actually Landsteiner's student) announced a fourth blood group, type AB. Its sera could cause both types A and B to clump.

Scientists now differentiate the blood groups by the presence or absence on the red blood cells of two antigens, called A and B, and their associated antibodies, alpha and beta, found in the plasma. Group A blood has the A antigen and the beta antibody that fights the B antigen, and group B has the B antigen and the alpha antibody that fights the A antigen. Group AB contains both A and B antigens, but neither antibody; and group O blood has neither antigen but both antibodies.

Landsteiner's work didn't immediately lead to the practical application of safer human-to-human transfusions, but within a couple of decades, medical professionals would begin the routine type-matching of blood. This helped to ensure that their patients could avoid the adverse, and often fatal, reactions to transfused blood of the wrong type. Landsteiner received the Nobel Prize in 1930 for his discoveries.

Reuben Ottenberg

In 1907, shortly after physician Ludvig Hektoen (1863–1951) suggested that type-matching would be a useful screening measure in transfusions, another physician, Reuben Ottenberg (1882–1959), put it to the test at Mount Sinai Hospital in New York. He repeatedly and successfully tested his assumptions. He also declared that type O can be transfused to anyone regardless of their blood type, because it contains no antigens and therefore does not agglutinize in response to antibodies. In 1912, physician Roger Lee affirmed Ottenberg's assumption that people with type O blood are "universal donors," and extended the idea to note that a person with type AB blood is a "universal recipient" and can safely accept donations of any blood type.

With this work, blood-type incompatibility was no longer a danger in blood transfusions. Transfusions were still not completely without risk, however, because the medical community still was unaware of the Rh factor.

Alexis Carrel

Another clinical advancement occurred in the early 1900s. Alexis Carrel (1873–1944), a French surgeon, worked on new techniques for sewing severed blood vessels together. He also expanded the technique in 1908 to transfusions by suturing donor arteries to recipient veins for direct blood transfer. Though the latter proved highly impractical, his work was quite useful for linking blood vessels during organ transplantations. He won the Nobel Prize in Physiology or Medicine in 1912 for his efforts, and continued his line of research, eventually co-inventing an "artificial heart" with famed aviator Charles Lindbergh that could temporarily oxygenate organs that were awaiting transplantation.

Luis Agote, Albert Hustin, and Richard Lewisohn

As the medical community was implementing cross-matching in blood transfusions, other researchers were trying to find ways to store blood so they could eliminate any need for direct transfusions. From 1914 to 1916, little more than a decade after the discovery of blood types, three researchers almost simultaneously were able to prolong blood storage. They were Luis Agote (1868–1954), Albert Hustin (1882–1967), and Richard Lewisohn (1875–1961). Agote and Hustin both found that when sodium citrate was mixed with blood, it would not clot. Lewisohn perfected the safe, effective dosage of sodium citrate per unit of blood.

Richard Weil, Francis Peyton Rous, and J. R. Turner

Richard Weil, Francis Peyton Rous (1879–1970), and J. R. Turner prolonged the utility of citrated blood. By 1917, Weil showed that the blood would keep for a few days through refrigeration, and Rous and Turner concocted a new citrate-glucose solution that extended the usable date to several weeks.

These advances paved the way for future blood banks, which provide medical facilities with a ready supply to meet demand. Blood banks would become particularly useful with the advent of World War I and the associated casualties.

Elliot Carr Cutler, Samuel Levine, and Henry Souttar

The 1920s were particularly notable for several contributions to heart care. Elliot Carr Cutler (1888–1947) and Samuel Levine, both American surgeons, and British surgeon Henry Souttar (1875–1964) began performing the first operations on defective heart valves. Cutler and Levine repaired a dysfunctional mitral valve by cutting off part of it, while Souttar simply poked his finger through a sealed mitral valve to reopen it.

Maude Elizabeth Abbott

Also in the 1920s, pathologist Maude Elizabeth Abbott (1869–1940) of the Woman's Medical College of Pennsylvania became well known for her work in categorizing and illustrating the congenital malformations of the heart combined with their physiological consequences.

Werner Forssmann

The first catheterization was performed in 1929 by a German surgical resident named Werner Forssmann (1904–1979). He felt that the best way to diagnose and treat cardiovascular conditions was from the inside. Apparently without the knowledge of his superiors, he slid a narrow tube, now called a cardiac **catheter**, into a vein at his elbow and guided it 25.6 inches (65 cm) through his vasculature to his heart. He had an x-ray taken of his chest to show the catheter in his right atrium and prove his accomplishment. Although he had dramatically established the safety of the procedure, which he felt would provide an excellent method of drug delivery, the medical community essentially branded him a crackpot. In 1956, however, his work was finally recognized by the Nobel Prize Committee. He shared the honor with André Frédéric Cournand (1895–1988) and Dickinson Richards (1895–1973), who in 1941 began employing the procedure as a diagnostic tool in a hospital setting.

Willem Kolff

Willem Kolff (b. 1911) created the first kidney machine, now known as a dialysis machine, in 1934 to take over the work of defective organs (see photo). The machine functioned by routing blood from the patient to a filtering system in the machine that removed waste products. The "clean" blood then flowed back into the patient. Kolff's machine was strictly a device for emergency, life-saving use. Eventually the dialysis machine was refined, and it is now used as an ongoing treatment option for kidney patients. Kolff is also known as an inventor of the artificial heart.

Willem Kolff (b. 1911) designed the first kidney machine, shown here, to filter waste products from the blood of patients with failing kidneys. © National Library of Medicine.

Ruth Darrow, Philip Levine, and R. E. Stetson

Three researchers set the stage for the discovery of the Rh factor, a protein on red blood cells that is present in most, but not all, people. They were physicians Ruth Darrow, Philip Levine (1900–1987), and R. E. Stetson. Darrow, of the Women's and Children's Hospital of Chicago, pored through the literature looking for family ties to the condition now known as Rh disease, which afflicts the fetus but not the mother. A meticulous examination of case studies led her to deduce in the late 1930s that a blood-borne antibody was transmitted from mother to the fetus through the placenta, and that antibody, while harmless to the mother, was dangerous to a fetus that had inherited the disorder from a parent.

By 1940, Levine and Stetson verified Darrow's contention and found the suspect antibody in the blood of a woman whose pregnancy had ended in a stillbirth.

Alexander Wiener

The Rh factor was discovered in 1940, almost four decades after Karl Landsteiner first identified the different blood groups. By studying red blood cells in rhesus monkeys, he, Alexander Wiener (1907–1976), and Philip Levine (who with R. E. Stetson had just identified the Rh antibody), identified a protein, which they called a Rhesus factor, and showed that approximately 80 percent of humans produce antibodies to it. The research also led to the conclusion that the fetus was stimulating the production of the antibody in the mother.

With this new information about Rh-positive and negative blood, cross-matching in blood supplies became safer, and the medical community now understood the relationship between mother and child in Rhesus disease. By the 1960s, Rh disease could not only be diagnosed, but also treated.

Charles Drew (1904–1950), who developed a method of separating out and then preserving blood plasma, opened the first blood bank in 1940. During World War II, he became medical director of the first American Red Cross blood bank. © National Library of Medicine.

Charles Drew

Another major advancement in blood transfusions came in 1940, when Charles Drew (1904–1950) of New York opened the first blood bank. Here, he would put to use his method of separating out and then preserving blood plasma, which he had already found to often be an adequate replacement for whole blood in transfusions. During World War II, he was named medical director of the first American Red Cross blood bank.

Edwin Cohn

As Drew was opening the first blood bank, physical chemistry professor Edwin Cohn (1892–1953) at Harvard Medical School was announcing a method for isolating proteins from liquid plasma. Through this process, called fractionation, he and his research group were able to separate out fibrinogen, gamma globulin, and albumin. The fraction containing albumin proved to be an especially useful clinical commodity during World War II to treat casualties. In 1951, Cohn's research team devised a tool for separating blood cells.

Paul Beeson

Despite the opening of blood banks, work on blood and plasma transfusions was far from over. In 1943, physician Paul Beeson (b. 1908) published findings in the *Journal of the American Medical Association* that related jaundice to transfusions, providing evidence that hepatitis could be transmitted through donated blood or plasma. Beeson is now considered a founder of the discipline of infectious diseases.

Alfred Blalock, Helen Taussig, and Vivien Thomas

Bypass surgery had its start in 1944 when Alfred Blalock (1899–1964) performed what is now known as the Blalock-Taussig operation, named for Blalock and Helen Taussig (1898–1986). Taussig ran the pediatric cardiac unit at Johns Hopkins Hospital, became interested in oxygen-deficient infants (so-called blue babies, a condition known as **cyanosis**), and began working with surgeon Blalock at Johns Hopkins Medical School on possible surgical solutions. Blalock, a vascular expert, and assistant Vivien Thomas (1910–1985) performed the first operation on a 15-month-old child. They were able to reroute blood past a blocked blood vessel and return adequate oxygen delivery, thus alleviating the cyanotic condition.

Wilfred Bigelow, Walton Lillehei, and John Lewis

In the mid- to late 1940s, Canadian surgeon Wilfred "Bill" Bigelow (b. 1913) opened the door for more complicated heart surgeries when he conducted experiments on animals to put them into an artificial hibernation of sorts. Because hibernating animals survive long periods in the winter with slow blood flow and low oxygen levels, he felt this same type of cooling would safely slow the human heart and give doctors enough time to perform heart surgery. At the time, doctors only had about four minutes to operate before the patient suffered brain damage. Bigelow was able to slow the animal heart rates to about one-ninth of normal and greatly increase the amount of time available for surgery.

Eight years later, surgeons C. Walton Lillehei (1919–1999) and F. John Lewis (1916–1993) used the cooling technique to perform open-heart surgery. They slowed the heart of a patient, a 5-year-old girl, for ten minutes—enough time to successfully correct a "hole" in her heart.

MID- TO LATE TWENTIETH CENTURY

John Gibbon

Bypass surgery leaped past the earlier accomplishments of John Blalock when American surgeon John Gibbon (1903–1973) used a heart-lung machine to completely reroute blood around the heart and lungs while he performed heart surgery in 1953 (see photo). The device, which he had developed over many years, artificially oxygenated the venous blood, then pumped it to the arteries. This first cardiopulmonary bypass lasted almost thirty minutes, and provided sufficient time for him to repair heart damage in an 18-year-old female patient.

Dennis Melrose

Two years after Gibbon's groundbreaking heart surgery in 1953, the research group of Dennis Melrose (b. 1921) in London employed potassium citrate and potassium chloride to stop completely the heart's movements in a dog that was hooked up to a heart-lung machine. On an unbeating heart, surgeons could make repairs that were difficult if not impossible when the organ was still slowly beating. The dog survived the heart stoppage. The work led to trials in other animals and in humans, and eventually became standard practice in open-heart surgeries.

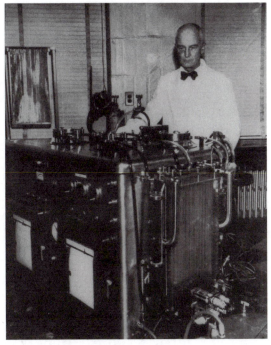

John Gibbon (1903–1973), inventor of the heart-lung machine. © National Library of Medicine.

Max Perutz

After about two decades of research at Keilin's Molteno Institute, Cavendish, and Cambridge University, organic chemist Max Perutz (1914–2002) in 1959 exposed the structure of the hemoglobin molecule using x-ray crystallography. Perutz won the 1962 Nobel Prize in Chemistry for his work. Co-recipient of the award was John Kendrew (1917–1997), another pioneer in determining protein structures.

Norman Shumway

American surgeon Norman Shumway (b. 1923) was intrigued by the possibilities of heart transplantation, and in 1961 took the heart of one dog and transplanted it into another. The second dog lived for three weeks, but his body finally rejected the new organ. Shumway's work showed that heart transplantation held promise if the problem of rejection could be surmounted.

Sir James Whyte Black

Common drugs used in heart ailments today come from a family of beta blockers that relax the heart muscle. Sir James Whyte Black (b. 1924) did the preliminary work in the 1950s and 1960s, and developed the first beta blocker, propanolol, in 1964. Beta blockers quickly made their way into the clinic for treatment of heart pain (angina), irregular heart beat, and high blood pressure. In 1988, Black received a Nobel Prize for his contributions.

Christiaan Barnard

South African surgeon Christiaan Barnard (1922–2001) became the first person to successfully perform a human-to-human heart transplant on December 3, 1967. The patient, Louis Washansky, survived for eighteen days. Barnard performed his second such procedure on Philip Blaiberg on January 1, 1968, and this patient lived about a year and seven months.

Robert Good

The first bone marrow transplant to cure immune deficiency disease occurred in 1968 when Robert Good (1922–2003) injected the marrow of a healthy woman into her ill brother. The first transplant between the two was largely unsuccessful, but a follow-up transplant worked, the marrow grew, and the man survived. Since then, bone-marrow transplantation has been refined and is used to treat difficult cases of leukemia, a cancer that over-produces white blood cells and makes too few red blood cells.

Tetsuzo Akutsu, Robert Jarvik, and William DeVries

In 1957, Willem Kolff and Tetsuzo Akutsu constructed an artificial heart and used it in a dog. (Recall that Kolff also invented the first kidney machine in 1934.) The dog survived for more than an hour. While others were attempting temporary artificial heart transplants in humans, Kolff continued to refine his device for long-term use and began working with Robert Jarvik (b. 1946). Jarvik eventually designed the Jarvik-7 model that was used in the first permanent, artificial-heart transplant in a human (see photo). At the University of Utah medical college, William DeVries (b. 1943) led the team of surgeons who implanted the device into patient Barney Clark on December 2, 1982. He died of complications nearly four months later, but the artificial heart never failed. One of the device's greatest detractions was that patients had to remain in the hospital and connected to life-maintaining machines.

Baruch Blumberg and Irving Millman

Although scientists and clinicians had known for nearly three decades that the liver inflammation known as hepatitis could be transmitted by blood and plasma tranfusions, it was not until 1971 that a researcher finally found out how to determine whether a blood sample was infected with the potentially fatal hepatitis B virus. Baruch Blumberg (b. 1925) of the National Institutes of Health was that researcher. He uncovered the specific part of the virus that initiated the body's immune reaction and the production of antibodies. With that information, Baruch and microbiologist Irving Millman (b. 1923) developed an assay to scan blood samples for the antibodies, and thus determine which donated blood supplies were contaminated. Blood banks started screening blood with the Baruch-Millman test in 1971. Later, the two men developed a vaccine against hepatitis B.

Much of the more recent cardiovascular history has been incorporated into current treatments for medical conditions, and will be covered in the next chapter.

Robert Jarvik (b. 1946), inventor of the Jarvik-7 artificial heart. The model, shown here, was used in the first permanent, artificial-heart transplant in a human. © AP/Wide World Photos.

Pacemakers Deliver a Steady Beat

Devices used to assist the heart to beat regularly, pacemakers entered the medical picture in the middle of the twentieth century. Their development is generally credited to Canadian electrical engineer John Hopps (1920–1998), Harvard researcher Paul Zoll (1911–1999), and American biomedical engineer Wilson Greatbach (b. 1919).

Through his work on hypothermia, Hopps found that he could resume the beating in a stopped heart by applying mechanical and electrical stimulation. In 1950, he developed a large, external pacemaker to maintain a regular heartbeat. Zoll built on the hypothermia work and designed a pacemaker with external electrodes positioned on a patient's chest. His device worked, but was unpractical and also caused skin burns. With his success, clinicians and researchers began to consider the possibility of an internal pacemaker.

With doctors William Chardack and Andrew Gage, Greatbach began work on a pacemaker small enough to be placed inside the body, and in 1958 he implanted the first internal pacemaker. Wires transmitted electric impulses directly to the heart through wires stitched to the muscle.

Pacemakers are still used today, although with some modification since Greatbach's invention, as effective treatment options for heart patients.

Unappetizing, but Effective: A Treatment for Pernicious Anemia

Investigations into the always-fatal disorder called pernicious anemia involved an unappetizing experiment that would eventually lead to a treatment . . . and a Nobel Prize.

The researchers were George Whipple, William Murphy, George Minot, and William Castle. Whipple (1878–1976) of Rochester University in New York discovered through experimentation on dogs in the 1920s that raw liver was an effective treatment for anemia that was due to blood loss. Murphy (1892–1987) and Minot (1885–1950) of Harvard University took an interest in Whipple's work, and in 1926 confirmed his results in humans, demonstrating that the ingestion of liver could treat pernicious anemia. Liver, it turned out, contains a good deal of vitamin B_{12}, which is necessary for the production of red blood cells.

It was Castle (1897–1990), one of Minot's students, who took the work further when he tested a hypothesis that the normal digestive system had an ability to take up vitamin B_{12} from the liver, but the digestive systems of patients with pernicious anemia lacked that ability. He fed lean beef, which has comparatively small amounts of vitamin B_{12}, to patients with pernicious anemia. They showed no improvement. He then fed lean beef to a healthy volunteer, and waited a while before passing a tube through the volunteer's nose and into his stomach to extract some of his stomach contents. Castle fed the gastric juice and partially digested meat to the patients with pernicious anemia. They improved. This showed that a healthy person's stomach contained something the patients' stomachs did not.

That "something" turned out to be a specific glycoprotein that permits the uptake of vitamin B_{12}. Through this work, Castle showed that pernicious anemia should be treated with both a source of vitamin B_{12} and the factor necessary for the body to use it. Of his research, Castle commented in 1929, "The foregoing experiments clearly demonstrate that the stomach contents of a normal man recovered during the digestion of a meal of beef muscle and subsequently incubated with additional hydrochloric acid contain a substance capable of causing remissions in certain cases of pernicious anemia comparable to those produced by moderate amounts of liver."

Despite the four researchers' contributions, only three shared the Nobel Prize for Physiology or Medicine in 1934 for their work: Whipple, Murphy, and Minot.

Cardiopulmonary Resuscitation

CPR, or **cardiopulmonary resuscitation**, is a well-known and often-taught first-aid measure to continue respiration and blood circulation in a patient whose heart has stopped beating. It involves compressions of the chest between the periodic delivery of oxygen to the unconscious patient.

The two parts of CPR—breathing and compressions—evolved from techniques that date back at least to surgeon William Tossach in 1732 and Edward Albert Sharpey-Schäfer in the early 1900s. Tossach used mouth-to-mouth breathing to revive a miner, and Schafer used chest compressions to trigger breathing.

In the 1930s, electrical engineer William Kouwenhoven (1886–1975) at Johns Hopkins was working on a research team to find a way to restore the heartbeat. Specifically, the team's intent was to find a way to revive power-company workers who had been electrocuted. Kouwenhoven developed the technique of regular chest compressions, also known as closed-chest cardiac massage, and found it successful in animal studies. The technique was apparently first tried on a human in 1958 by Henry Bahnson, who used it to revive a child. That success sparked the use of CPR as a staple in emergency care.

In addition, Kouwenhoven reasoned that if an electrical jolt could cause a person's heart to fibrillate, a second jolt might be able to return the heartbeat to normal. In 1933, he and his colleagues proved the hypothesis. By the 1950s, the team had developed a portable, closed-chest defibrillator that could jumpstart a nonbeating heart. This work brought cardiopulmonary resuscitation from the lab to the bedside, and the procedure and modern versions of the device are now staples of emergency care.

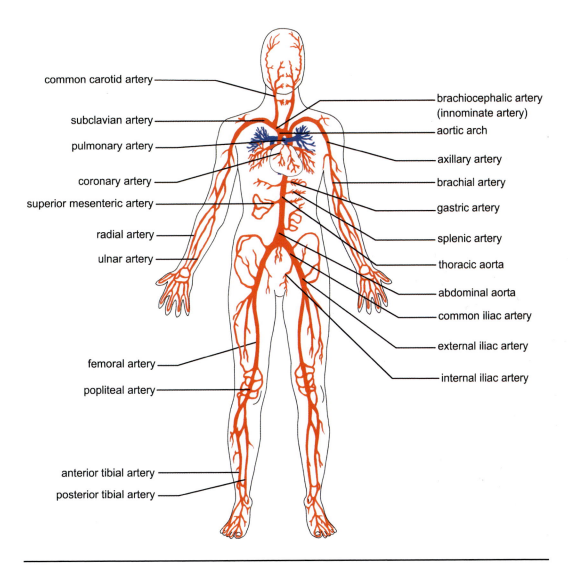

common carotid artery

subclavian artery

pulmonary artery

coronary artery

superior mesenteric artery

radial artery

ulnar artery

femoral artery

popliteal artery

anterior tibial artery

posterior tibial artery

brachiocephalic artery
(innominate artery)

aortic arch

axillary artery

brachial artery

gastric artery

splenic artery

thoracic aorta

abdominal aorta

common iliac artery

external iliac artery

internal iliac artery

Names of the major arteries of the systemic circulation.

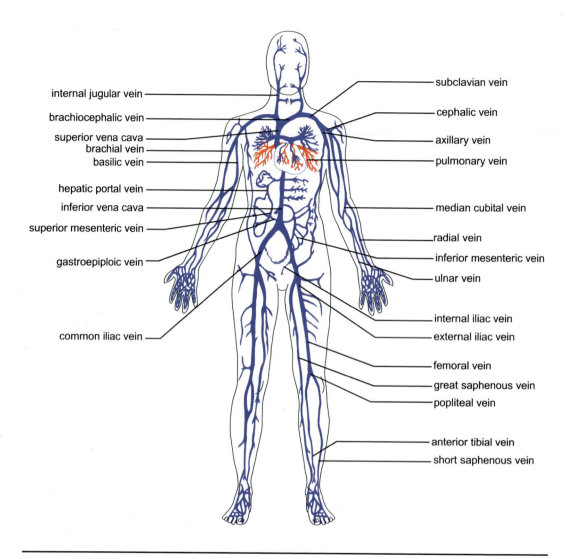

internal jugular vein

brachiocephalic vein

superior vena cava
brachial vein
basilic vein

hepatic portal vein

inferior vena cava

superior mesenteric vein

gastroepiploic vein

common iliac vein

subclavian vein

cephalic vein

axillary vein

pulmonary vein

median cubital vein

radial vein

inferior mesenteric vein

ulnar vein

internal iliac vein
external iliac vein

femoral vein
great saphenous vein
popliteal vein

anterior tibial vein
short saphenous vein

Names of the major veins of the human systemic circulation.

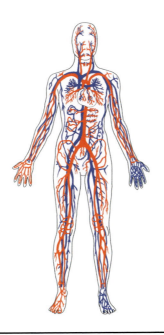

The major veins and arteries of the human circulatory system. Arterial vessels carry oxygenated blood, which is shown in red. Venous vessels carry deoxygenated blood, which is shown in blue. (Exceptions are the pulmonary arteries, which carry deoxygenated blood from the heart to the lungs, and the pulmonary veins, which deliver oxygenated blood from the lungs to the heart.) For the vessels of the extremities, the right hand and foot show the major arteries, and the left hand and foot reveal the major veins.

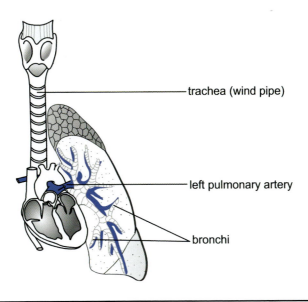

trachea (wind pipe)

left pulmonary artery

bronchi

Arteries of the pulmonary circulation. One pulmonary artery enters each lung, then branches into many arterial vessels to distribute deoxygenated blood to the alveoli.

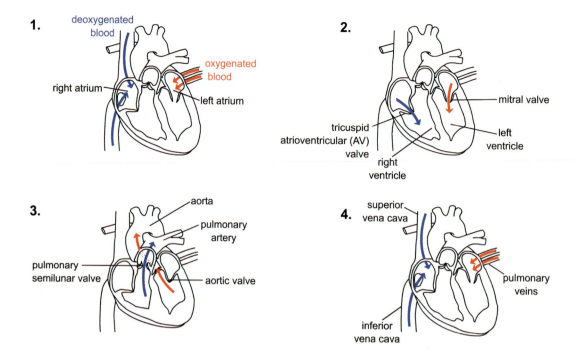

One cycle of a working heart. (1) The right atrium fills with deoxygenated blood flowing to the heart from the upper and lower body, while the left atrium fills with newly oxygenated blood from the lungs. (2) The atria contract, and heart valves open to allow blood to move from the left atrium to the left ventricle, and the right atrium to the right ventricle. Once the atria are empty, the valves swing shut. (3) The ventricles contract, forcing the blood through exit valves and to the pulmonary artery on the right side and the aorta on the left. (4) Another cycle begins as blood enters the atria. (Note that the pulmonary arteries are different from other arteries in that they carry deoxygenated blood.)

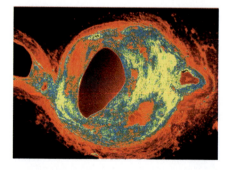

Atherosclerosis, also known as hardening of the arteries, results when cholesterol, fat, and other materials—collectively called plaque—build up along artery walls and obstruct the flow of blood. In this colorized cross-section of a human artery, plaque has narrowed the vessel's opening to less than a fifth of its normal size. © Custom Medical Stock Photo.

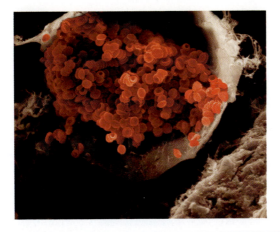

This image provides some indication of the size and number of red blood cells that pass through a single blood vessel. © Michael Webb/Visuals Unlimited.

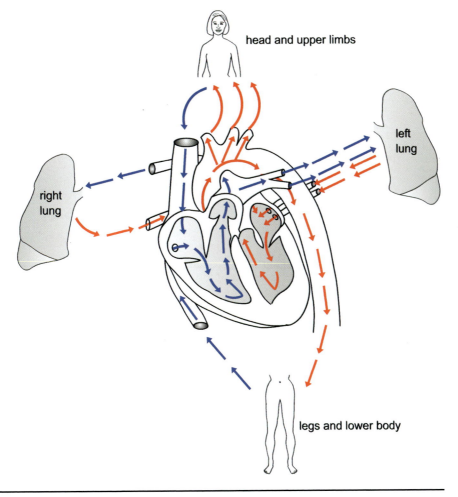

The direction of blood flow through the heart and to the rest of the body.

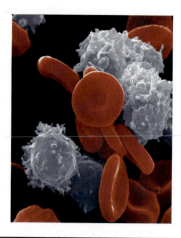

Red and white blood cells. A single drop of blood from the average person contains 4.5–5.5 million red blood cells. © Dr. David Phillips/Visuals Unlimited.

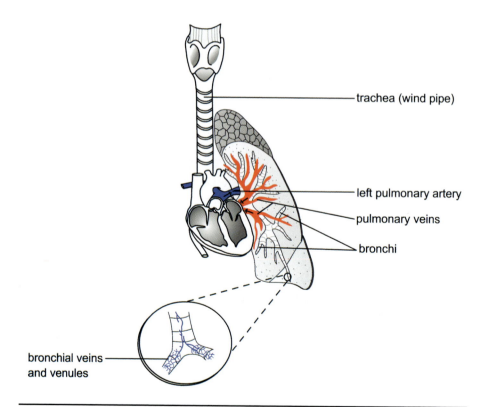

trachea (wind pipe)

left pulmonary artery

pulmonary veins

bronchi

bronchial veins
and venules

Veins of the pulmonary circulation. Newly oxygenated blood from the lungs collects in the pulmonary veins for transport to the heart, where it is pumped to the rest of the body. The lungs also have many bronchial veins, which drain blood from the bronchi and a portion of the lungs, eventually delivering the blood to the superior vena cava, which brings blood into the heart's right atrium.

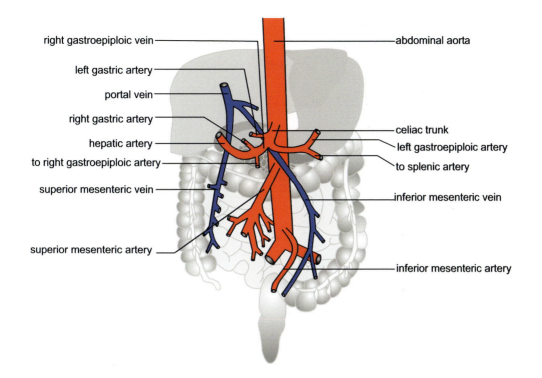

right gastroepiploic vein ——— abdominal aorta

left gastric artery ———

portal vein ———

right gastric artery ——— ——— celiac trunk

hepatic artery ——— ——— left gastroepiploic artery

to right gastroepiploic artery ——— ——— to splenic artery

superior mesenteric vein ——— ——— inferior mesenteric vein

superior mesenteric artery ——— ——— inferior mesenteric artery

Major arteries and veins of the hepatic and digestive systems.

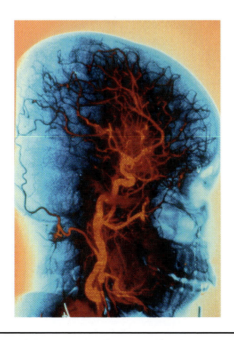

Color-digitized angiogram of the human head, showing blood vessels. An angiogram is an x-ray representation of the blood vessels. © Custom Medical Stock Photo.

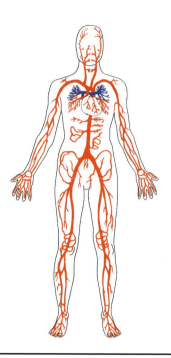

The major arteries of the human circulatory system. Arteries carry oxygenated blood, shown in red. Exceptions are the pulmonary arteries, which deliver deoxygenated blood from the heart to the lungs.

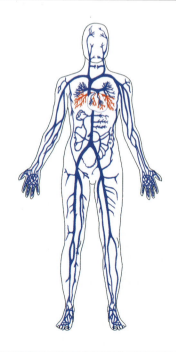

The major veins of the human circulatory system. Veins carry deoxygenated blood, shown in blue. Exceptions are the pulmonary veins (shown in red), which deliver oxygenated blood from the lungs to the heart.

Cardiovascular Disorders and Diseases

Considering the intricacies of the vascular bed, the nonstop action of the heart, and the sheer volume of blood that must travel around the body every day, it's not surprising that the cardiovascular system does not always work perfectly in everyone. This chapter will describe many of the ailments, problems, diseases, and other disorders associated with the heart, blood, or vessels. Because treatments can vary from doctor to doctor and even day to day, the information provided here should be used for educational purposes only and not taken as clinical advice. Each of the conditions listed in this chapter will include a definition of the disorder, associated symptoms, causes, and some of the common treatment options. Information on lifestyle changes that can improve cardiovascular health are listed in Chapter 11.

ACQUIRED IMMUNODEFICIENCY SYNDROME

Acquired immunodeficiency syndrome, or **AIDS**, is a state of severely depressed immune function, in which the body has little defense against disease, including everyday infections. (See "History of AIDS.") Symptoms vary widely, depending on which infections the patient experiences.

HIV, or **human immunodeficiency virus**, assails the immune system, specifically targeting white blood cells known as CD4 T cells. When the body begins to succumb to disease it can no longer fight, a person is said to have AIDS. A person can often live for many years before an HIV infection progresses to AIDS. The virus can spread from person to person through

History of AIDS

At the end of the 1970s and beginning of the 1980s, the medical community began to hear reports of a disorder affecting mainly homosexual men. Originally called gay cancer, and then gay compromise syndrome or gay-related immunodeficiency disease (GRID), the name was changed to acquired immune deficiency syndrome, or AIDS, in 1982 when it began to appear among people who had received blood transfusions. All patients with the disease experienced an immune system failure. In a *Newsweek* article, an official from the Centers for Disease Control in 1982 commented, "When it began turning up in children and transfusion recipients, that was a turning point in terms of public perception. Up until then it was entirely a gay epidemic, and it was easy for the average person to say 'So what?' Now everyone could relate."

By 1983–1984, researchers had isolated and identified the AIDS virus, known as human immunodeficiency virus (HIV), and in 1985 screens were developed to test blood for virus antibodies. In the United States alone, the number of people dying from complications of AIDS had already risen from 159 in 1981 to 2,122 in 1983 and to 12,592 in 1985, the year that U.S. movie star Rock Hudson died from complications of the disease.

By 1987, when the death toll topped 41,000 in the United States, the U.S. Food and Drug Administration approved **AZT (zidovudine)** as an anti-HIV treatment. Little and/or inaccurate information precipitated a public scare that AIDS could be transmitted through simple human contact (or less), leading to widespread biases and sometimes violence against AIDS patients. As the U.S. surgeon general attempted to allay fears in this country, researchers continued to pursue answers through medications, including Intron A and Roferon A, to treat an HIV-induced cancer called Kaposi's sarcoma, and others to treat conditions relating to the patients' immune collapse.

Through the 1990s, education about AIDS improved and the public began to understand that AIDS is transmitted mainly by unprotected sex, dirty hypodermic needles, and tainted blood supplies used in transfusions. At the same time, however, HIV was spreading rapidly and had reached epidemic numbers, particularly in developing nations where the infection rate can range from 20 to 40 percent of the population. In the world overall, the number of infected people in 2002 had reached 42 million. Added to that, nearly 22 million people had already died of complications from the disease.

On the positive side, many countries have dropped their infection rates. Uganda, for example, has used educational campaigns to help lower its HIV infection rate from about 14 percent in the early 1990s to 5 percent in 2002.

Advances in treatment have also occurred. Physicians now select from a battery of medications to fight HIV spread in a patient's body. While the drugs have been successful in prolonging an infected individual's lifespan from a couple of years to decades, no cure exists, and the disease continues to be infectious throughout the person's lifespan.

a variety of vectors, including transfusions with infected blood. For this reason, many people with hemophilia (see separate section) contracted the virus before the medical community understood the origin of AIDS and began screening blood.

Since the late 1970s and early 1980s when AIDS was first recognized by the medical community, treatments have markedly improved and are now allowing HIV-infected individuals to live longer and healthier lives. Physicians now have a battery of medications, called antiretroviral drugs, that fight HIV by shutting down the enzyme-based pathway that the virus requires either to replicate or to spread. These drugs do not cure HIV infection, but can slow its progress. Side effects of the typically multidrug therapy are common, however, and can be severe.

ANEMIA

Patients with **anemia** have too few red blood cells or too little hemoglobin in their circulating blood, leading to insufficient oxygen delivery to body tissues. Often, anemia takes the form of **iron deficiency anemia**, and is particularly common in women who lose blood each month during the menstrual cycle. Iron is required to make the hemoglobin molecules that carry oxygen in the blood, so an iron deficiency affects oxygen delivery. **Aplastic anemia** is an uncommon but particularly dangerous type of anemia in which the individual's bone marrow is unable to produce blood cells. Only one in ten patients survive the malady. In **hemolytic anemia**, red blood cells are destroyed much faster than they are replaced, resulting in an overall loss of erythrocytes. People who have **pernicious anemia** are unable to absorb vitamin B_{12}, which is necessary for the production of red blood cells. In **thalassemia**, the patient is unable to properly synthesize hemoglobin. See the separate section on sickle-cell anemia later in this chapter. Depending on the type and severity of the anemia, a patient's symptoms may include:

- Ashen skin, mucus membranes
- Weakness and/or fatigue
- Dizziness
- Difficulty breathing
- Headache and/or slight fever
- Rapid pulse
- Soreness of the mouth and tongue
- Diarrhea and/or nausea
- Numbness and/or tingling in the hands and feet
- Angina pectoris (see separate section later in this chapter)

- In women, cessation of menstruation

- In the severe form of thalassemia, enlarged liver, spleen (see section on splenomegaly later in this chapter), and heart

Low red blood cell counts can result from a number of conditions that cause the loss or destruction of red blood cells, or the improper or inadequate formation of red blood cells. Iron-deficiency anemia results when iron intake or absorption is insufficient. Pregnancy and heavy menstrual flow can exacerbate the iron deficiency. Other causes of blood loss, including stomach ulcers, can bring about iron deficiencies. Aplastic anemia can be caused by pregnancy, some cancer treatments, as well as severe infections and a number of drugs. Hemolytic anemia may result from genetic and other diseases as well as toxins, and may affect newborns. A lack of vitamin B_{12} metabolism is the culprit in pernicious anemia. Individuals with pernicious anemia either ingest insufficient B_{12} in their regular diets or cannot process B_{12} because they are unable to make the protein required for its metabolism. B_{12} is required to produce red blood cells in the bone marrow, so a deficiency results in a low erythrocyte count. The inability to process B_{12} can result from diabetes (see separate entry later in this chapter) or other diseases or surgery to remove part of the stomach. Persons more than 60 years old are also more prone to the condition than younger individuals. Thalassemia is a genetic disorder.

Iron-deficiency anemia is usually treated with dietary changes and/or iron supplements. For aplastic anemia, the treatment usually involves repeated blood transfusions, and possibly a bone marrow transplant. A common treatment for hereditary hemolytic anemia is the surgical removal of the spleen, the site where the destruction of many red blood cells occurs. Patients with pernicious anemia receive injections of B_{12}. The treatment for the severe form of thalassemia is continued blood transfusions. The mild form of thalassemia typically has no symptoms and requires no treatment.

ANGINA PECTORIS

Temporary chest pain or pressure that radiates from the heart to the shoulder and left arm, or sometimes from the heart to the abdomen, is known as **angina pectoris**. Angina attacks are short lived and may continue from a few seconds to a few minutes. Common symptoms of angina include:

- Acute, steady pain radiating from the heart through the shoulder and to the left arm, and occasionally from the heart to the abdomen

- A feeling of high anxiety or impending doom

- Sweating

- Pale, sometimes cyanotic face

- Labored breathing
- Racing pulse

Angina occurs when the heart muscles receive insufficient oxygen, typically from inadequate blood flow. The condition often arises from atherosclerosis (see separate section later in this chapter), in which narrowed arteries impede the blood flow.

Drugs are commonly issued for patients with angina pectoris. These may be beta blockers and calcium antagonists that lower the heart's demand for oxygen, or coronary **vasodilators** like nitroglycerin that relax and therefore widen blood vessels in the heart. Atherectomy, balloon **angioplasty**, laser angioplasty, stents, and bypass surgery are other options (see the section on heart attack later in this chapter). Sometimes, medical professionals recommend a procedure known as transmyocardial revascularization, which makes dozens of miniscule holes in the ventricle and allows blood to reach cardiac tissue that previously was receiving too little oxygen. The procedure relieves the pain associated with angina.

AORTIC DISSECTION

Also known as **aortic aneurysm**, an **aortic dissection** is bleeding from the aorta due to a tear in the vessel's lining. The tears usually occur in the thoracic portion of the aorta and result in bleeding into the aorta wall, sometimes leading to a blood-flow stoppage, heart failure, or rupture of the aorta. Patients usually encounter several indications of an aortic dissection, and most quite suddenly. They may encompass:

- Severe, often sharp chest pain that may radiate to the back, shoulders, arms, jaw, and sometimes to the abdomen
- Dizziness, confusion, and sometimes fainting
- Pale, sometimes clammy skin
- Rapid, sometimes weak pulse
- Drop in blood pressure
- Sweating
- Inability to move the extremities
- Nausea, perhaps vomiting
- Difficulty breathing, especially when lying down, sometimes accompanied by coughing

Aortic dissection may be the result of a blunt-force trauma to the chest that tears the aorta. Even in the absence of such an injury, individuals may experience a tear or a ballooning and bursting of the aorta. The cause of these non–injury-related dissections is often associated with atherosclerosis, high blood pressure (see separate sections later in this chapter), and congenital or other disorders that cause a weakening of the aorta. Research presented at the 2002 Scientific Sessions of the American Heart Association reported that cocaine use is also a contributing factor to this condition.

Once the problem is properly diagnosed, which can be difficult due to its similarities to other cardiovascular conditions, medical professionals may recommend emergency surgery to close the tear or to replace the damaged section of the vessel. Stents may also be prescribed to hold open the aorta and resume blood flow. In addition, patients may receive a number of drugs to reduce pain and moderate blood pressure.

ARRHYTHMIA

Heart **arrhythmias** are heart rhythms that deviate from the normal "lubb-dupp" pattern and pace of the heart. Arrhythmias include fast beating, or **tachycardia**; slow beating, or **bradycardia**; lack of beating, or **asystole**; and other abnormal patterns, like ventricular contraction before complete filling of the chamber. Persons with heart arrhythmias often are unaware of the condition. Those who report symptoms typically have one or more of the following:

- Weakness, sometimes fatigue
- Dizziness, lightheadedness, and rarely fainting
- Rapid, irregular pulse (in tachycardia)

The common cause of heart arrhythmia is a problem with the electrical system that initiates each beat of the heart. It may arise in the atria, the ventricles, or the electrical connection between the two.

Depending on the type of arrhythmia, different treatments apply. For tachycardia and abnormal patterns, for instance, beta blockers (described under the section on heart attack later in this chapter) are used to encourage the heart to beat slower and less forcefully. Calcium channel blockers (see the section on high blood pressure later in this chapter) slow the heart rate, but also relax blood vessels to reduce demands on the heart. Other commonly used agents include **digoxin** to retard electrical conduction in the heart and slow its beat, and **digitalis** that dampens rapid beating of the atria, a condition known as **atrial fibrillation**. For bradycardia or asystole, **atropine** or **isoproterenol** may be prescribed to activate the heart muscle.

Beyond pharmaceutical agents, physicians may recommend that a **defibrillator** be implanted. This device detects arrhythmias and delivers a mild electrical jolt that is just powerful enough to put the heart back on a regular pace. A cardiac pacemaker is another option. This device delivers an electrical impulse to artificially stimulate a normal heartbeat, especially in patients with bradycardia. In emergency situations where the patient's heart flutters uncontrollably and blood flow nearly ceases, medical professionals may use a different type of defibrillator, which delivers a strong electrical pulse to the heart in an attempt to shock it back into a normal sinus rhythm. This procedure is also known as **electrical cardioversion**.

For persons whose heart has stopped beating, cardiopulmonary resuscitation is a life-saving technique.

ATHEROSCLEROSIS

Also known as **hardening of the arteries**, **atherosclerosis** is a narrowing of arterial walls. The deposits, collectively called **plaque**, create rough, irregular surfaces that are prone to blood clots (see color insert). Patients are generally oblivious to the condition unless they experience complications due to blood clots or severe reductions in blood flow to the heart or to the brain, which can result in heart attack, stroke (see separate sections later in this chapter), or other disorders.

Scientists believe atherosclerosis may begin when the endothelium of the arteries is damaged by various means, including high levels of circulating cholesterol (see separate section on high cholesterol later in this chapter) and fat in the blood or high blood pressure that puts stress on arterial walls. People who smoke, are obese, or have diabetes (see separate entry later in this chapter) are also prone to this condition. Endothelial damage appears to spur cholesterol, fats, and other materials to deposit upon and line arterial walls, narrowing the lumen and hindering blood flow.

Because many heart attacks result from clots in narrowed arteries, the treatment options for atherosclerosis are similar to those for heart attack. They may include medication to prevent blood clots, angioplasty and stents to open narrowed arteries, and bypass surgery to reroute blood past particularly clogged arteries.

An intriguing report issued in 1997 indicated that men who donate blood are less likely to develop atherosclerosis and therefore heart disease. Published in the August issue of *Heart*, the study noted that men who were donors slashed their risk of heart disease by up to 30 percent. Researchers hypothesized that the donations were lowering the men's blood levels of iron, which some believe is related to atherosclerosis.

CARBON MONOXIDE POISONING

Carbon monoxide poisoning can occur when a person is exposed to this colorless, odorless gas. Carbon monoxide is composed of one atom each of carbon and oxygen. Normally, hemoglobin in the blood delivers oxygen (molecules comprising two atoms of oxygen) to the body tissues and removes carbon dioxide (molecules made of one atom of carbon and two of oxygen). When a person breathes carbon monoxide, the hemoglobin preferentially picks up and delivers this molecule instead of oxygen, and if the exposure continues, cells begin to die. Prolonged exposure can be fatal. Carbon monoxide poisoning is characterized by:

- Headache
- Drowsiness
- Dizziness and/or confusion
- Weakness
- Nausea and/or vomiting
- Chest pain
- Unconsciousness and possibly death if exposure is prolonged

Improperly functioning furnaces, water heaters, and other gas appliances are common sources of carbon monoxide. People may also be poisoned if they are in a confined space with a fuel-burning device. For example, a person in a garage with a running vehicle or in a travel trailer with a space heater or camp stove may be exposed to dangerous levels of carbon monoxide. People with carbon monoxide poisoning may receive oxygen, perhaps hyperbaric (high-pressure) oxygen, to replace carbon monoxide in the system with oxygen.

CARDIOMYOPATHY

Cardiomyopathy is a disease of the heart muscle that makes it weak and unable to function effectively. Several varieties exist. **Dilated cardiomyopathy**, also known as congestive cardiomyopathy, is a condition that leads to an enlargement of one or more heart chambers. In **hypertrophic cardiomyopathy**, the heart walls are thickened, which decreases the volume of the chambers, particularly the left ventricle, and affects the heart's ability to pump effectively. In **restrictive cardiomyopathy**, abnormal tissue causes the ventricles to stiffen, which affects their ability to adequately pump blood. Persons with cardiomyopathy typically complain of:

- Weakness and/or fatigue
- Fluid accumulation in the extremities, sometimes in the lungs
- Breathing difficulties, particularly when lying down or during physical activity
- Fainting spells

- Nausea, especially in restrictive cardiomyopathy
- Arrhythmias (see separate section earlier in this chapter)

The cause of dilated cardiomyopathy is unknown. Hypertrophic cardiomyopathy is a genetic disorder. The cause of restrictive cardiomyopathy is typically one of a number of ailments, including the inflammatory condition called sarcoidosis or the genetic condition known as hemochromatosis that leads to iron deposits throughout the body.

Dilated cardiomyopathy has no cure, but symptoms may be treated with various medications, including beta blockers, calcium channel blockers, vasodilators, and digitalis to ease the heart's workload; diuretics to alleviate fluid accumulation; and antiarrhythmia agents to regulate heartbeat. Occasionally, a left ventricular assist device or even a heart transplant is recommended (see the section on heart failure later in this chapter). For hypertrophic cardiomyopathy, doctors may prescribe a range of drugs similar to that for dilated cardiomyopathy. In addition, they may recommend a pacemaker to set a rhythm that is easier on the heart, as well as surgery to trim back the thickened septum. Patients with restrictive cardiomyopathy may receive diuretics and possibly other medications to treat the underlying condition at the root of the cardiomyopathy.

CAVERNOUS ANGIOMA

Cavernous angiomas are tiny, leaky pouches filled with blood. The cause of this condition remains unknown, but the pouches arise from abnormal blood vessels. These reddish accumulations of vessels can be too small to see, as large as a peach, or anywhere in between. An individual may have one angioma or several. The bleeding from an angioma can be dangerous, depending on the location of the angioma and the amount of blood loss. Often, a patient is unaware he or she has an angioma. Symptoms vary depending on the site of the angioma, and are often intermittent. They may include:

- Seizure, possibly multiple seizures
- Weakness in the extremities
- Vision problems
- Balance problems
- Hemorrhage
- Headache
- Stroke-related symptoms (see section on stroke later in this chapter)

Often small angiomas present no symptoms and disappear on their own. Recurring hemorrhages within the same angioma, even if it is small, may require medical intervention, including surgery to remove it. Angiomas in the brain can be particularly dangerous as blood either spills out into the brain or builds up inside the now-growing angioma. These are usually surgically removed. Patients who experience seizures or headaches may receive medications for those symptoms.

CHURG-STRAUSS SYNDROME

Churg-Strauss syndrome is an inflammation of the blood vessels that can lead to organ failure and sometimes death. Other types of vessel inflammation, called vasculitis, exist. Patients with this syndrome typically experience one or more of the following:

- Fever

- Weight loss

- Numbness or weakness in the legs and arms, sometimes with fatigue

- Difficulty breathing, cough, and/or chest pain

- Small bumps, particularly on the hands and feet

- Sometimes seizures or confusion if the brain is affected

- Inflammation of the nasal passage and/or sinus in asthmatic patients

The cause of this rare syndrome remains unknown, but it is found in individuals who have asthma or allergies. Scientists have noted that the body's immune system makes extra white blood cells, especially eosinophils, in patients with this condition. Patients usually receive medication to treat the inflammation combined with immune-suppressing drugs.

CIRRHOSIS OF THE LIVER

Cirrhosis of the liver is a chronic degenerative condition resulting in substantial liver damage that causes scarring and eventually affects the organ's function. **Biliary cirrhosis** is one form of the disease and is characterized by inflammation of bile ducts. Scar tissue resulting from cirrhosis serves to block normal blood flow through the liver, thus impairing its ability remove toxins from the blood. People with cirrhosis typically do not realize they have the disease until it has advanced. Complaints at that time may include:

- Enlargement of the liver at first, followed sometimes by severe shrinking

- Enlargement of the spleen (see the section on splenomegaly later in this chapter)

- Itching

- Fatigue

- Loss of appetite and/or nausea
- Thin, red, so-called spider veins on the skin, especially noticeable on the face and nose
- Swelling of the extremities and/or the abdomen
- In men, atrophied testicles and swollen, sometimes sore, breasts
- **Jaundice**, especially in biliary cirrhosis
- In biliary cirrhosis, itching and/or formation of small, yellow bumps on the eyelids, hands, and elbows

Alcoholism is a common cause of cirrhosis, and is likely responsible for at least half of all cirrhosis-related deaths in North America. In the system of an alcoholic, the liver receives the alcohol from the digestive system, eventually turning it into molecules called cytokines that damage tissues, beginning with the liver itself. As the liver damage continues, the organ's function is affected. After alcoholism, chronic hepatitis is the next most common cause of cirrhosis. Here, the liver cells become inflamed and damaged, and function poorly. In addition, some patients develop the disorder from autoimmune diseases, some of which may stem from viral infections, or from other conditions, such as cystic fibrosis.

One of the first lines of treatment for this condition is cessation of alcohol consumption if that is determined to be the cause. For those whose cirrhosis is caused by hepatitis, medical professionals may prescribe drugs, especially some **interferons**, to reduce liver damage or mount an more effective offense on the hepatitis viruses. (See the separate section on hepatitis later in this chapter.) Patients may also receive medications or other treatments for symptoms like fatigue, itching, or internal bleeding. Occasionally, physicians will recommend a liver transplant if the disease is advanced.

COARCTATION OF THE AORTA

Coarctation of the aorta is one of a number of common congenital heart disorders, and involves a narrowing of part of this vessel, the largest artery in the body. Sometimes the condition presents no symptoms. However, physicians typically identify coarctation of the aorta by one or more of the following:

- High blood pressure, particularly in the arms
- A very slight pulse in the legs
- Heart murmur (see section on valvular abnormalities later in this chapter)

Severe narrowing of the aorta can lead to heart failure (see separate section later in this chapter).

Repair of the narrowed artery is a common treatment. This may be accomplished with balloon angioplasty (see the section on heart attack later in this chapter), surgical removal of the narrowed section of vessel, or placement of a stent to hold open the vessel.

CONGENITAL AFIBRINOGENEMIA

Individuals with the inherited blood disorder called **congenital afibrinogenemia** either lack plasma fibrinogen, which is one of the proteins that participates in blood clotting, or have defective plasma fibrinogen that is unable to function in blood clotting. The blood in persons with this disorder cannot coagulate. Patients generally only experience unstoppable bleeding following an injury or in surgery, and, sometimes for women, in childbirth. The amount of bleeding varies with the severity of the trauma. Only rarely do patients experience spontaneous bleeding.

Treatment involves cessation of the bleeding. This is usually accomplished with transfusions of plasma that contains fibrinogen and/or concentrated fibrinogen.

CORONARY HEART DISEASE

Coronary heart disease is an umbrella term used to describe many heart-related conditions, all of which involve narrowed coronary arteries that affect the blood flow to the heart. In the United States, at least 12 million people have coronary heart disease. Because symptoms are often absent, this number may be lower than the actual total. People with early coronary heart disease typically have no symptoms, making this a particularly dangerous disease. By the time symptoms arise, the narrowing of the arteries is usually fairly advanced. Those reporting symptoms generally complain of one or more of the following:

- Discomfort, dull ache, or sharp pain in the chest, particularly when doing strenuous exercise
- Discomfort in the shoulder, throat, or left arm
- Shortness of breath

Together, these three symptoms are collectively grouped under the condition known as angina (see separate entry earlier in this chapter). Extreme narrowing can also result in **heart failure** or **heart attack** (see separate entries later in this chapter).

Coronary arteries typically narrow when deposits of fat, cholesterol, and other materials build up along the inside of arteries. The buildup, known as plaque, not only narrows the lumen of the blood vessel, but also hardens in a process called "hardening of the arteries," which is known as atherosclerosis (see the separate section earlier in this chapter). If plaque becomes extreme, it can severely affect blood flow.

A recent study in the April 24, 2001, issue of *Circulation* reported an increased risk for heart disease among African Americans who harbor a genetic mutation that may lead to an increase in blood clots. The mutation affects the gene that is responsible for producing a protein called thrombomodulin. This protein blocks clots. Those with the mutation don't make the protein and are thus more susceptible to clot formation.

Treatment is similar to that for heart attack. (See the section on heart attack later in this chapter.) In addition, researchers are now investigating the possibility of using stem cells to make heart muscle cells. Stem cells are undifferentiated cells from which all other types of cells arise, including heart cells. If scientists can encourage these stem cells to develop into heart cells, they could be enormously useful in repairing damaged heart tissue. Already, at least one group of scientists from the Technion-Israel Institute of Technology in Haifa reported that they were able to spur embryonic stem cells to become cardiomyocytes. They published their work in the August 2001 issue of *Journal of Clinical Investigation.*

DIABETES

The bodies of people with **diabetes** either produce too little of the hormone called insulin or they do not use it properly. Insulin, which is secreted by the pancreas, allows the body cells to use the energy, specifically the sugar called glucose, that our bodies derive from food. Among diabetics, the pancreas either doesn't make enough insulin or the cells don't respond to it correctly. Diabetic or not, however, the bloodstream continues its normal job of accepting glucose from the digestive system, so glucose accumulates in the blood and winds up as a waste product that exits the body in the urine without having nourished the cells. The heightened sugar level can also damage the heart and blood vessels. Patients with diabetes often experience decreased blood flow, which can present particularly acute problems in the feet. This decrease, often called peripheral vascular disease, can result in slow healing times for foot injuries, sometimes leading to foot ulcers.

Diabetes comes in three main forms: type 1, type 2, and **gestational diabetes**.

The most common form of diabetes is type 2, also known as **diabetes mellitus** or adult-onset diabetes. Historically the typical patient was more than 55 years old and overweight, but the number of affected children has greatly increased in the past decade as the number of obese individuals has risen. In type 2 diabetes, the pancreas makes insulin as it should, but the cells don't respond to insulin's signal to take up glucose.

Type 1 diabetes, sometimes called **juvenile diabetes**, usually develops by the time a person reaches 25 years old. It is an autoimmune disease in which the immune system attacks part of the body. In this case, the immune system mistakenly views the insulin-producing cells in the pancreas as foreign entities and begins to destroy them, leaving the pancreas with little capacity to make insulin.

Gestational diabetes is a transient form of the disease that strikes pregnant women, then disappears following birth. According to the National Institute of Diabetes and Digestive and Kidney Diseases (NIDDK), women who have had gestational diabetes have a higher risk than average for developing type 2 diabetes later on.

The NIDDK reports that perhaps half of all people with diabetes are unaware that they have the disease. Those who know they have diabetes mellitus may have many symptoms, which are typically more severe among people with type 1 disease. They include:

- Retinopathy, or bleeding capillaries in the retina of the eye, which can cause blindness

- Blurry vision

- **Hyperglycemia**, or elevated blood-sugar levels

- Increased urination and thirst

- Itchy skin

- Constant hunger, but frequently accompanied by weight loss that may lead to weakness

- In severe cases, headache, nausea and/or vomiting, and difficulty breathing

- Extreme fatigue

- Frequent infections, including yeast infections in women

- Slow healing from even minor injuries

According to a study published in the July 18, 2001, issue of the *Journal of the American Medical Association*, inflammation may also be an early symptom of adult-onset diabetes in women. The U.S. study revealed that women who showed evidence of low-grade inflammations were up to 4.2 times more likely to develop diabetes than women without such evidence.

Diabetes can have long-term effects, including a slow erosion of bodily functions that may result in blindness, heart failure, stroke, and other organ failures. The increased likelihood of infections may lead to gangrene and amputation. Researchers suspect that genetics and perhaps a viral infection may be involved in type 1, and note that persons of certain ethnicities or from certain families, and those who are obese, are more likely to develop the type 2 form of the disease.

A daily injection of insulin combined with a restrictive diet and an exercise program make up the typical treatment for individuals with type 1 diabetes. Occasionally, patients may receive pancreas transplants, which can cure them of the disease. Patients with type 2 and mild symptoms may be placed on a diet and exercise program, while those with more severe symptoms may require insulin supplementation. Recent studies indicate that ACE inhibitors (see the section on heart attack later in this chapter) may help to prevent kidney damage in diabetics.

Two 2001 studies indicated that a drug called p277 lessened the need for insulin injections in individuals newly diagnosed with type 1 diabetes. The drug, which is made from protein fragments, limits the destruction of cells, called beta cells, that make insulin. The findings were released in the November 24, 2001, issue of *Lancet*, and at the September 2001 meeting of the European Association for the Study of Diabetes.

In other diabetes-related work, researchers in 2003 developed a new drug treatment that combats the dizziness often associated with diabetes and other conditions. A report on the drug was published in the September 2003 issue of *Journal of Neurology and Neurosurgical Psychiatry*. The drug, called pyridostigmine, regulates the dips in blood pressure that occur when a person shifts from lying down to standing up. The drops, called baroreflexes, are often amplified in diabetics. Pyridostigmine assists the communication of nerves that govern the baroreflexes.

Physicians stress self-care for diabetes, insisting that patients carefully monitor their blood-sugar levels. If the level rises too high in the condition known as hyperglycemia, or too low in the condition called **hypoglycemia**, the individual may rapidly become confused, faint, and even die if emergency care is not delivered immediately. New research reported at the 2001 Scientific Sessions of the American Diabetes Association indicated that hypoglycemia can also impair cognitive ability.

ECTOPIC PACEMAKER

An ectopic pacemaker is any set of heart cells that take over the typical pacemaker function. Normally, the sinoatrial node (SA node) or the pacemaker serve as the initiation point for the heart's electrical system. Located in the atrial wall near the entrance of the superior vena cava, these small and weakly contractile, modified muscle cells fire spontaneously to deliver regular electrical impulses and maintain the heartbeat. In addition to the SA node, the heart also has backup regions that can take over beat initiation if the pacemaker is compromised. These alternative sites are called ectopic pacemakers. Occasionally an ectopic pacemaker may cause tachycardia in the atria, but not the ventricles, and sometimes cause bradycardia (see the section on arrhythmias earlier in this chapter).

Ectopic pacemakers may become active when certain areas of the myocardium receive inadequate blood flow and/or too little oxygen. They are often associated with coronary artery disease, heart attack, heart failure, such hormonal conditions as hyperthyroidism, low extracellular potassium levels, and use of diuretics or chemotherapeutic agents. If ectopic pacemakers result in irregular heartbeat, physicians may prescribe medication like that used for arrhythmia to deal with this condition. Generally, however, the "treatment" centers around identifying and correcting the problem that diverted pacemaker activities from the SA node to another site.

ENDOCARDITIS

Endocarditis is a sometimes-fatal bacterial infection of the endocardium, which is the membrane lining the heart. Typically the infection remains in the lining of the valves, but may reach into the lining of the atria and ventricles. Symptoms vary, but may include:

- Fever
- Weakness
- Difficulty breathing
- Headache
- Loss of appetite
- Tachycardia and sometimes development of a heart murmur (see sections on arrhythmia and valvular abnormalities in this chapter)
- Joint pain

The cause is not always determined, but infective endocarditis is a common complication of rheumatic fever (see the separate entry later in this chapter), which can damage heart valves. It may also arise from congenital heart disease (see the entry on coronary heart disease earlier in this chapter).

A regimen of antibiotics may be prescribed, combined with bed rest. For advanced cases, medical professionals may recommend prosthetic heart valves to replace the damaged ones, or shunts and catheters to improve blood flow in vessels that have been affected by the infection. A small study reported in the March 20, 2001, issue of *Circulation*, indicated that the breast-cancer drug tamoxifen may be useful in widening arteries in men with heart disease.

ESSENTIAL THROMBOCYTHEMIA

Essential thrombocythemia is a disorder that results in an abundance of blood platelets in the bone marrow. In essential thrombocythemia, a single megakaryocyte mutates, then rapidly reproduces, and the quickly growing population of "super megakaryocytes" makes an excess of platelets. Other disorders that allow bone marrow cells to reproduce in excess are **polycythemia vera** and **myelofibrosis**, which affect erythrocytes, megakaryocytes, and/or fibroblasts.

The disorder is usually asymptomatic, but when they occur, symptoms may include:

- Itchy skin
- Splenomegaly (see the separate section later in this chapter)
- Blood clots

Essential thrombocythemia is the result of a mutation in the genetic material of a bone marrow cell. The mutation causes it to proliferate wildly. Scientists are unsure of its origin, but suspect it may have a genetic basis and possibly be initiated by radiation exposure. Patients may undergo plateletpheresis, in which the platelets are removed from blood drawn from the patient, then transfused back into the patient. Chemotherapy drugs are also sometimes prescribed.

HEART ATTACK

When people think of health problems associated with the cardiovascular system, heart attacks usually come to mind first. A heart attack, also known as a **myocardial infarction,** happens when the supply of oxygen to a portion of the heart muscle is curtailed to such a degree that the tissue dies or sustains permanent damage. Approximately 1.1–1.5 million heart attacks occur in the United States each year.

Patients who have had heart attacks frequently report a combination of symptoms, although some people recount no symptoms and only find that they have had an attack during a routine physical checkup. (One study indicated that fully a third of patients report no chest pain.) Commonly reported symptoms include:

- Crushing chest pain, as if a huge anvil has been set on the breastbone, or **sternum**
- Abdominal pain similar to severe heartburn

- A radiating upper back pain, which is frequently reported by women in lieu of chest pain

- Sharp, long-lasting pain that spreads from the chest to typically the left shoulder and arm, or through the neck and jaw

- Sudden shortness of breath for no apparent reason, perhaps accompanied by a cough, and sometimes exacerbated when lying down

- Dizziness and/or fainting

- Nausea and/or vomiting

- Sweating for no apparent reason

- Rapid heartbeat (see the separate section on arrhythmia earlier in this chapter)

- Feeling of impending doom and/or anxiety

- Lasting fatigue

Generally, a heart attack is the result of a stationary blood clot, or **thrombus**, that suddenly blocks one of the large coronary arteries that supply the heart. The clot, which can close off part or all of the blood flow, results in a sufficient blood stoppage so that some cardiac muscle cells suffer oxygen deficiency and die or sustain enough damage so that they are no longer able to contract. Many blood clots are a result of atherosclerosis, a condition that results when plaque (deposits of fat, cholesterol, and other materials) builds up inside arteries, effectively narrowing the lumen. Plaque that tears or breaks can trigger a blood clot to form. Because the artery is already narrower at the site, the clot has the potential to severely or completely block blood flow. (See the separate entry on atherosclerosis earlier in this chapter for additional information.)

Risk factors for heart attacks include high blood pressure (see the separate entry later in this chapter), a diet rich in fats, obesity, diabetes (see the separate entry earlier in this chapter), and high cholesterol (see the separate entry later in this chapter). In addition, heart attacks are more common in men than women, in older persons, and in those who have a family history of heart attacks. A study published in a June 2003 supplement to *Environmental Health Perspectives* demonstrated a link between air pollution and heart attack. By investigating the types of air pollution present when more than 5,000 people suffered heart attacks, they were able to show that days of high air pollutants correlated with a greater number of heart attacks.

In rare cases, a heart attack can be caused by a "spasm" in a blood vessel, which pinches off blood supply to the heart. The cause of these spasms are unknown.

Treatment for heart attacks has changed dramatically over the past few decades, and may include angioplasty and stents, bypass surgery, and/or various drugs. Physicians also may prescribe dietary alterations, exercise, and other lifestyle changes.

Angioplasty is a procedure in which the medical practitioner inserts a catheter through the blood vessels and to the narrowed artery that helped cause the heart attack. The end of this catheter contains some type of device to widen the vessel. Often, the device is a balloon that expands and compresses the plaque against the artery wall. This widens the lumen, and blood can flow more freely. In a similar procedure, called **coronary atherectomy**, the catheter uses a device that actually cuts through the plaque to reopen the lumen. **Laser angioplasty** uses lasers at the end of the catheter to clear plaque. These procedures may also be employed as preventive measures against future heart attacks. Catheters can also be used to install **stents**, which are tiny, tube-shaped metal devices that remain in place at the problem site to hold the artery open. Arteries sometimes reclose even after angioplasty or stent placement. Reports in 2003 indicated that stents laced with anticlotting drugs may help to prevent future narrowing.

In addition, medical professionals use catheters to pinpoint the location of the narrowed artery. In this procedure, called **coronary angiography**, a catheter is used to deliver a dye into the heart's arteries. As a result, cardiologists have a clear view of blood flow in the arteries and can discern which artery has the "pinched" lumen. Other diagnostic tools that may be used by hospitals include **echocardiography** to take an ultrasound of the moving heart, **nuclear** or **radionuclide ventriculography** to view the heart and major blood vessels, and measurements of various telltale chemicals in the blood. These chemicals, like **myoglobin** and **troponin**, are released by dead cardiac muscle cells. Recent research reported in a 2002 issue of *Circulation* revealed that persons who suffered a fatal heart attack also showed an increased concentration of a chemical called C-reactive protein in their blood. The researchers contend that this chemical may also be a good marker for potential heart attacks.

While angioplasty is very effective when the person has only one or two narrowed arteries, it is not practical when more arteries are obstructed. When this occurs or when angioplasty proves unsuccessful, the physician may recommend **coronary artery bypass surgery** (also known as **coronary artery bypass grafting**). Instead of repairing the affected region, this procedure reroutes blood around it. The surgeon begins by removing a piece of vessel from elsewhere in the body, often a leg or arm vein. Next, he or she opens the chest and stops the heart, clips the affected artery above and below the narrowing, and reattaches it using the piece of vessel. The blood "bypasses" the narrowing.

Some of the medications typically used in heart-attack treatment include nitrates, thrombolytic agents, beta blockers, and ACE inhibitors. **Nitrates**, such as **nitroglycerin**, dilate blood vessels, which makes it easier for blood to flow and not only lowers blood pressure and eases the demands on the heart but also reduces pain associated with the heart attack. **Thrombolytic agents** fall into the family of so-called clot-busting drugs. These drugs, such as streptokinase, urokinase, antistreplase, and tissue plasminogen activator (see the section on stroke later in this chapter), employ different methods of doing the same thing: breaking up blood clots. Like nitrates, **beta blockers** and **angiotensin-converting enzyme inhibitors** or **ACE inhibitors**, both reduce the demands on the heart. Angiotensin-converting enzyme (ACE) is necessary to make angiotensin II, a chemical that narrows blood vessels. The ACE inhibitors block the enzyme, allowing vessels to relax and consequently decreasing blood pressure. Medications like ACE inhibitors are collectively known as vasodilators, because they dilate blood vessels. Beta blockers also reduce the heart's workload, but in a different manner. Beta blockers inhibit adrenaline and other so-called beta-adrenergic substances that trigger heart muscle action. When these substances are blocked, the heart beats slower and less forcefully.

Patients who have had heart attacks are commonly placed on continuing medication, which may range from blood thinners and aspirin to prevent new clots from forming, to drugs that prevent erratic heartbeat (see section on arrhythmia earlier in this chapter). The prescribed drugs are also almost always accompanied by recommended lifestyle changes (see Chapter 11).

HEART DISEASE

Heart disease is a blanket term used to describe many heart-related conditions, including such disorders as valve disease, congenital disease, heart arrhythmias, and coronary heart disease. These conditions are treated separately in this chapter.

HEART FAILURE

When the heart no longer can carry out its pumping function adequately, the condition is known as heart failure. The heart contractions become so weak that they no longer pump sufficient blood from the ventricles. As a result, blood circulation slows, cells become poorly oxygenated, and the veins begin to hold more blood. **Congestive heart failure** is heart failure accompanied by fluid in the lungs and extremities. Each year, about 300,000 Americans are newly diagnosed with heart failure. Typical symptoms include:

- Dry cough or shortness of breath, especially when lying down
- Fatigue
- Dizziness or fainting
- Swollen ankles or legs

- Fluid in the lungs
- Increased urination

Heart failure often results from a heart attack (see separate entry), and is typically treated with medications, including nitrates, ACE inhibitors, diuretics, and inotropic agents. Descriptions of nitrates and ACE inhibitors are included in the preceding section on heart attack. **Diuretics** trigger the kidneys to remove more water and sodium from the plasma, thereby reducing swelling (also called **edema**) in the extremities and also lowering the total blood volume. The latter reduces the heart's workload by decreasing the amount of blood it has to pump. Digitalis and inotropic agents are medications used to boost the force of the heart's contractions. This allows the heart to push blood more forcefully, which increases blood pressure enough to ensure that oxygenated blood is reaching all of the body's tissues.

Angioplasty (see the preceding section on heart attack) is another option if the physician finds that a narrowed artery is drastically restricting blood flow to the heart. Occasionally, a medical professional may recommend that a patient undergo an operation to treat heart failure if the condition is caused by a faulty valve, a severe arrhythmia (see the section on arrhythmia earlier in this chapter), or some other surgically correctible problem. Although **heart transplants** are possible, physicians rarely recommend this option and only if all other options have been exhausted. Typically, patients receive a temporary mechanical implant, called a **left-ventricular assist device**, to help the heart pump blood until a compatible donor heart is located. Recent research published in the November 15, 2001, issue of the *New England Journal of Medicine* noted that the devices are also useful to extend the lives of terminally ill patients who are ineligible for transplants. The devices work by routing blood from the left ventricle to the device, which then pumps the blood into the patient's aorta. On average, patients who had the device lived more than eight months longer than those who didn't.

HEMOGLOBIN C DISEASE

In **hemoglobin C disease**, an inherited disorder, patients have abnormal hemoglobin that crystallizes in red blood cells, deforming them and making them targets for destruction by the spleen. Patients with this disease, which mainly affects persons of African descent, often report no symptoms. Those who experience problems generally complain of:

- Joint pain
- Splenomegaly (see the separate section later in this chapter)
- Thin, brownish, so-called angioid streaks beneath retinal vessels

- Dark-colored ("pigmented") gallstones
- Mild anemia (see the separate section earlier in this chapter)

Patients generally receive treatment for the symptoms, including regular eye examinations to monitor angioid streaks, which can affect vision. Medications or surgery may be prescribed to deal with problematic gallstones. The care options for anemia and splenomegaly are listed under the associated entries in this chapter.

HEMOPHILIA

In **hemophilia**, a bleeding disorder, the proteins for blood clotting either are lacking or function improperly. A number of these proteins, called factors, are all required to control bleeding. Two of the most common forms of this disease are hemophilia A and hemophilia B. Those with the A type lack factor VIII, and those with type B lack factor IX. Von Willebrand's disease is similar to hemophilia, and is treated in this chapter under its own entry. Indications of this disease vary widely, particularly between men and women. They are:

- Nosebleeds
- Bruising even from minor bumps
- Excessive bleeding following injury or surgery, and, in severe hemophilia, from even very minor cuts
- In women, excessive bleeding during menstruation or childbirth
- Anemia (see the separate section earlier in this chapter)
- Joint swelling and pain, which can be extreme

Hemophilia is genetic in origin. Although it is typically considered a male disease, both males and females can have the disease. Often, however, females carry the trait and pass it on to their children, but have milder symptoms, if any.

Patients with mild or moderate hemophilia require minimal medical intervention and only following injury or at times of surgery. That intervention is usually injections of the missing factor in synthetic form. Those with severe hemophilia typically require preventive factor supplements, sometimes needing injections every week. These treatments can prevent much of the extreme pain associated with the disease. Researchers are also studying whether gene therapy may be useful in treating hemophilia. In the June 9, 2001, issue of *The New England Journal of Medicine*, for instance, a group of researchers from Harvard Medical School and from the Massachusetts-based Transkaryotic Therapies inserted the gene for factor VIII into plasmids (rings of bacterial DNA) and introduced them to study participants with hemophilia A. Although it wasn't successful in all study participants, some began to produce the factor in sufficient quantities that they required no other care for blood-clotting for nearly a year.

HEMORRHAGE

Hemorrhage is the loss of blood, and usually refers to a significant amount of blood loss in a short time. The loss of more than two pints (0.9 liters) of blood over a short span usually leads to vessel collapse because there is simply insufficient blood to keep all of the vessels filled. With too little blood, venous flow slows and the circulation eventually stops. Within a couple of minutes, brain cells begin to perish, followed shortly thereafter by body cells. Victims of nonfatal hemorrhage typically experience:

- Dizziness, lightheadedness, weakness, and/or fainting
- Confusion
- A rapid, feeble pulse rate
- Rapid respiration
- Pale, clammy skin, sometimes cyanotic, especially at the lips and finger-nails
- Blood in the stool, urine, and/or vomit

External or internal injuries are common sources of hemorrhage, both of which occur following injury to blood vessels. External bleeding is typically the result of a wound that breaks the skin. Internal bleeding can result from traumatic injury, disease, or other damage to organs and tissues.

Different types of hemorrhage require different treatments. In cases of external bleeding, the blood loss must be stopped. This is usually accomplished by applying direct pressure to the wound, although pressure to the nearest upstream artery is sometimes required. More severe cuts may demand stitches or surgery to curtail the flow, and blood transfusions may be necessary. The care for internal hemorrhage varies widely based on the cause of the blood loss.

HEPATITIS

Hepatitis is an inflammation of the liver, usually due to viral infection. Different types of hepatitis result in varying forms and severity of symptoms. These may include:

- Swelling and tenderness of the abdomen near the liver
- Muscle and/or joint pain

- Fatigue

- Mild fever, sometimes accompanied by nausea and/or vomiting

- Loss of appetite

- Occasionally diarrhea and/or dark-colored urine

- Jaundice (yellowing) and/or itching of the skin

Hepatitis usually occurs as a result of viral infection, but the inflammation may also be due to alcohol and/or drug use, or chemicals. It may take one of five forms: A, B, C, D, or E. Hepatitis A and E viruses spread through fecal contamination of food or water, sometimes via raw shellfish or dirty utensils (see photo). The hepatitis B virus can be transmitted in several ways. Transmission may occur through sexual contact, by blood transfused from tainted supplies, by dirty hypodermic needles contaminated with the virus, by close and repeated physical contact as might occur between siblings in a family home, or at birth or shortly thereafter from the mother to the newborn. Hepatitis C is usually spread directly from one person to another through blood or contaminated needles. Hepatitis D is spread in a similar manner to hepatitis C, but it only occurs if the person already has hepatitis B.

Patients usually recover in a few weeks to six months or more, depending on the type of hepatitis. Recovery is generally complete for all but those with hepatitis C, which often becomes chronic, sometimes leading to cirrhosis of the liver (see the separate section earlier in this chapter). Treatments for patients with severe symptoms may include drugs like interferon alpha-2b. Vaccines are available for hepatitis A, B, and E. Because hepatitis D only occurs after an individual already has hepatitis B, the vaccines also prevent hepatitis D.

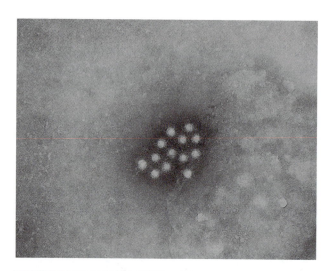

An electron micrograph of the hepatitis A virus. One of five types of hepatitis virus, hepatitis A virus spreads through fecal contamination of food or water. © Centers for Disease Control and Prevention.

HEREDITARY HEMORRHAGIC TELANGIECTASIA

Hereditary hemorrhagic telangiectasia (HHT), also called **Osler-Weber-Rendu syndrome**, is similar enough to hemophilia that it was only in the late 1800s that HHT was finally recognized as something distinct from hemophilia (see the separate section earlier in this chapter). Instead of the blood-clotting problem indicative of hemophiliacs, individuals with HHT lack capillaries, so blood moves directly from the arterial system to the venous system. Normally, blood is under a higher blood pressure in the arterial system, but the pressure declines when the blood enters the capillaries. This is because the blood from each vessel branches out into many capillaries, much as a water current eases as a river splits into a number of smaller creeks. The pressure gradient between the arterial and venous system is high enough to cause ruptures where the two systems meet, and this results in bleeding. Usually, arteriole-to-venule connections rupture and bleed at the skin surface, including the mucus membranes of the nose. Ruptures of artery-to-vein connections are more likely to occur in organs, particularly the stomach, intestines, and brain. Liver and lung ruptures are also relatively common in people with HHT. Because ruptures and bleeding may occur in different places in individual patients, the symptoms vary. In general, however, patients experience at least a few of the following:

- Recurring nosebleeds, which may last a minute with minimal loss of blood or may last several hours with enough blood loss to require a transfusion

- Reddish to purplish spots or small, netlike patches of fine red vessels on the skin, usually on the lips, on the tongue, and in the nose

- Blood in the stools, which may appear black

- Anemia (see separate section earlier in this chapter)

- Weakness, and sometimes fatigue and dizziness

- Difficulty breathing

- Chest or back pain, sometimes accompanied by numbness or loss of function in an extremity

- If untreated, stroke or heart failure

Treatment usually involves care for the individual ruptures, and varies depending on the site affected. Mild external bleeding is usually treated with in-home remedies, but extreme external bleeding and internal bleeding typically requires medical intervention. A patient with a long-lasting nosebleed may receive a transfusion to make up for the lost blood. For a patient who has repeated nosebleeds that become disruptive to his or her daily life, a physician may recommend laser therapy to seal off at-risk vessels or, if that fails, a skin graft to replace the more delicate nose lining with thicker skin. A patient who has problems with skin lesions may similarly receive laser therapy to relieve ruptures and bleeding. The treatment for HHT-induced anemia is the same as that included under the section on anemia earlier in this chapter.

Therapy may include an aggressive search for potential rupture sites in the brain and lungs, followed by various techniques to remove them.

HIGH BLOOD PRESSURE

Also called hypertension, high blood pressure occurs when the pressure of the blood against the vessel walls exceeds normal limits. An estimated 50 million Americans have high blood pressure, although many have no symptoms and are unaware of it. Another 45 million fall into the category of prehypertension, which makes them twice as likely to develop high blood pressure as they age. Patients typically report no symptoms, although a few mention headaches or dizziness. Hypertension, however, is a primary risk factor in heart attack, heart failure, and **stroke** (see separate sections elsewhere in this chapter).

High blood pressure results from a narrowing of the arterioles, which, in turn, increases blood pressure. High blood pressure causes the heart to work harder to pump blood through the narrowed vessels. People with sustained high blood pressure risk an enlarged heart that arises from the increased workload, heart attack, stroke, and damage to the kidneys.

For most people with high blood pressure, the cause is mostly unknown, although genetics and lifestyle are suspected to play a role. In addition, the following people are more likely to have hypertension: men, African Americans, overweight individuals, people more than 60 years old, women using oral contraceptives, diabetics, smokers, and heavy consumers of alcohol.

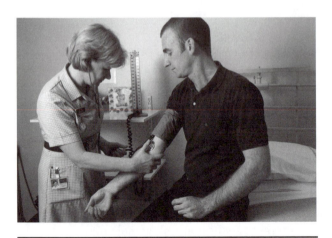

People with some kidney, neuronal, and hormonal disorders often have high blood pressure, but this accounts for only about 5 percent of all hypertension. In addition, hypertension is a fairly common side effect of pregnancy, as well as of numerous medications.

Recent research also links hypertension to a bacterial infection. According to a study published in the February 1998 issue of *Hypertension*, the microbe that causes pneumonia and bronchitis appears to be associated with hypertension. The bacterium, called *Chlamydia pneumoniae*, is also suspected

A nurse takes a patient's blood pressure with a blood-pressure cuff, also known as a sphygmomanometer. In the average resting adult, the blood pressure in the aorta rises to 120 mmHg following a heartbeat and then drops off to 80 mmHg before the next heartbeat. © Mediscan/Visuals Unlimited.

as a participant in the incidence of heart disease and stroke. Conducted by Heartland Hospital in Birmingham, England, the study reviewed nearly 246 men. Half had high blood pressure, and the other half didn't. Of those who had hypertension, 35 percent also had *Chlamydia* infection. Among the control group, only 18 percent showed the infection.

Medical professionals usually prescribe dietary changes, an exercise regimen, and other lifestyle alterations if hypertension is newly diagnosed. For long-term hypertension, physicians may prescribe medications, such as diuretics, ACE inhibitors, or beta blockers (described under the section on heart attack earlier in this chapter). Another common medication is a **calcium channel blocker**. Calcium channel blockers inhibit calcium, which blood vessels require for contraction. With lower levels of calcium, the blood vessel walls are more relaxed, allowing blood to flow more freely.

HIGH CHOLESTEROL

High cholesterol, or hypercholesterolemia, occurs when a person has too much cholesterol in the bloodstream. **Cholesterol** is a naturally occurring fat-like substance that the body uses to make hormones, bile, and other materials. It also occurs in nerve tissue in the brain and in cellular membranes throughout the body. When a person has excess cholesterol, however, it can begin to line the arteries, which leads to atherosclerosis and possibly other cardiovascular problems, including heart attack (see separate sections in this chapter). Cholesterol actually comes in two forms. One is **low-density lipoprotein (LDL)**, which is often called the "bad" cholesterol. This is the type of cholesterol that builds up on blood-vessel walls. The other form is called **high-density lipoprotein (HDL)**, the so-called good cholesterol, and curtails LDL accumulation in vessels. A person is considered to have high cholesterol if the total cholesterol level (HDL and LDL) equals 240 mg of cholesterol/dL of blood or more. A borderline-high diagnosis is issued for total cholesterol levels between 200 and 239 mg/dL.

High cholesterol imparts no symptoms, which is why medical professionals urge everyone to know their own cholesterol levels and act on them, if necessary. Genetics and age play primary roles in an individual's cholesterol level. Cholesterol typically rises with increasing age. In addition to these factors, diet and weight can influence cholesterol levels.

To lower cholesterol levels, physicians usually first recommend a program of diet and exercise. If these are unsuccessful, they may prescribe cholesterol-lowering drugs. Options may include one or a combination of several drugs, such as **statins, bile acid sequestrants, nicotinic acid, gemfibrozil, probucol,** and **clofibrate.** Research on statins, which have become increasingly popular, suggests that the drugs work by relaxing blood vessels and improving their flexibility. They accomplish this feat, according to an article in the July 20, 1998, issue of *Circulation*, by increasing the amount of a chemical, called nitric oxide, that regulates blood vessels.

HYPERKALEMIA

Hyperkalemia is a high concentration of the mineral potassium in the circulating blood. Patients generally have no symptoms, but when symptoms are present, they may include:

- Irregular heartbeat, often bradycardia (see separate section on arrhythmia earlier in this chapter)
- Abnormal pulse, typically slow and weak
- Nausea

The kidneys are the primary site for maintenance of potassium levels, and kidney disorders can lead to hyperkalemia. Specifically, hyperkalemia can arise from any disorder that disrupts the kidneys' ability to remove excess potassium from the body. These disorders may include inflammations of the kidneys or the glomeruli (clusters of capillaries in the kidneys). The problem is often exacerbated by burns, surgery, tumors, and any other tissue trauma that triggers an increased excretion of potassium from the cells into the extracellular fluid and bloodstream.

Patients may receive emergency calcium to temporarily correct arrhythmias, as well as glucose and insulin to boost potassium uptake by tissues, thus alleviating the bloodstream overload. Low-potassium diets are typically recommended. Diuretics and other medications may be used to increase potassium excretion. When all else fails, physicians may recommend dialysis to take over kidney function.

HYPOKALEMIA

Hypokalemia is a low concentration of the mineral potassium in the circulating blood. Patients may experience:

- Muscular weakness, possibly paralysis
- Intermittent spasms or cramping
- Gastrointestinal problems, including constipation and abdominal cramps
- Nausea and/or vomiting
- Frequent urination and excessive thirst
- Sudden drops in blood pressure upon standing up

Gastrointestinal disorders or renal disease are two common causes for hypokalemia. In both cases, too much potassium is removed from the blood.

Sometimes the deficiency results from an inadequate diet or from diuretics and other medications that affect potassium levels. Potassium supplements are often all that is required to treat mild cases. For more severe cases, the patients may receive intravenous potassium, as well as care for specific symptoms.

IMMUNE THROMBOCYTOPENIC PURPURA

In the blood disorder called **immune thrombocytopenic purpura (ITP)**, antibodies attack and destroy the platelets that are necessary for blood clotting. This leads to bleeding problems. A temporary form of the disease is more common in children, usually completely disappearing within a year. Another, chronic form of the disorder is primarily seen in adults. Individuals with immune thrombocytopenic purpura may experience the following:

- Bruising
- **Petechiae,** which are small, reddish-purple spots on the skin, in the nose, inside the mouth, and in other mucus membranes
- Nosebleeds
- Bleeding from the gums, in the gastrointestinal or urinary systems, and/or in the brain

Certain viral infections and immune disorders can bring about the disorder. The use of some drugs, as well as pregnancy, may also be involved, but often the cause of ITP remains unknown. Some patients have mild symptoms and require no intervention. When needed, medical care for this disorder usually involves removal of the cause, if possible, such as a discontinuation of particular drug. In more severe cases, physicians may prescribe **corticosteroids** and/or immune globulin to counter the body's autoimmune response.

LEUKEMIA

Leukemia is a blanket term applied to various cancers of the white blood cells. In these cancers, the body makes an overabundance of abnormal white blood cells that are incapable of fighting infections, while making too few red blood cells and platelets. Leukemia may progress slowly (chronic) or quickly (acute), and different types are differentiated by the types of white blood cells affected. Some of the more common types include chronic lymphocytic leukemia, and acute or chronic myeloid leukemia. Often-reported symptoms associated with leukemia include:

- Weakness and fatigue
- Anemia (see separate section earlier in this chapter)

- Fever and chills
- Frequent infection
- Loss of appetite and weight loss
- Swollen or tender lymph nodes, liver, and/or spleen
- Bleeding and bruising for even the slightest of injuries
- Sweating, especially at night
- Bone and/or joint pain
- Swollen, bleeding gums
- Ashen skin, sometimes with small, purplish spots (a condition called petechiae)
- In acute leukemia, also headache, vomiting, confusion, and/or seizures

Genetics may be related to the incidence of leukemia, but its cause is currently unknown. The disease is more common in men than in women, and among certain ethnic groups. Some scientists suspect that exposure to high-energy radiation, electromagnetic fields, or various chemicals may also lead to leukemia in some patients.

Chemotherapy, sometimes combined with radiation therapy, is commonly prescribed to kill abnormal cells, although treatment is highly specialized depending on the patient and the type of leukemia. For those patients who do not respond to chemotherapy, bone-marrow transplants may be recommended. In this procedure, the patient's own bone marrow—the source of the abnormal cells—is destroyed and then replaced with marrow from a healthy donor. When successful, the donated marrow grows and begins to function normally, effectively curing the patient.

LYME DISEASE

Lyme disease, which dates back to 1975 and an outbreak in Lyme, Connecticut, is spread by a bite from a tick. This can become a debilitating disease, leading to painful arthritis and facial paralysis. Early symptoms are typically mild and often unnoticed, but later manifestations can be quite severe. Disease characteristics may include:

- Round, red, bull's-eye rash that expands in size over several days
- Flu-like symptoms, such as fever, chills, and fatigue
- Muscle and joint pain
- Swollen glands
- Painful arthritis, sometimes with swelling

- Numbness
- Bell's palsy, which is facial paralysis typically on one side
- Meningitis
- Occasionally heart arrhythmias and/or myocarditis

Lyme disease is caused by an infection of the bacterium *Borrelia burgdorferi*. Some ticks transmit the bacterium to humans. Diagnosed patients typically receive antibiotics to fight the bacterial infection. If treated early, recovery is rapid and complete. If treated months or even years after the initial infection, repeated antibiotic treatments may be required. Sometimes patients experience permanent joint or nervous damage.

MONONUCLEOSIS

Often shortened to "mono," **mononucleosis** occurs when the circulating blood has an abundance of abnormal, mononuclear white blood cells. People with mononucleosis generally experience several of the following:

- Fever
- Sore throat, often with visible, white patches in the throat
- Swollen, frequently tender lymph nodes
- Enlarged spleen (see section on splenomegaly later in this chapter) that may rupture in severe cases
- Fatigue
- Loss of appetite

A ruptured spleen carries its own set of symptoms, including abdominal pain on the left side, dizziness, and difficulty breathing.

Infection with the Epstein-Barr virus causes mononucleosis, and this virus can be passed through saliva or mucus. Among teens, it is often called the "kissing disease," but it can also be spread through other means, including the cough of an infected person. Mononucleosis typically runs its course over about a month, during which time home remedies and over-the-counter painkillers are commonly recommended to treat most symptoms.

MULTIPLE SCLEROSIS

Multiple sclerosis (MS) is a chronic disease in which the immune system mistakes the insulation around nerve fibers in the central nervous system, including the brain and spinal cord, for foreign material. It then begins sending out T cells and antibodies to destroy the "foreign material." Without this insulation, called the **myelin sheath**, the nerves short-circuit, eventually resulting in scar tissue, which is called a plaque or a lesion, that can no longer conduct electrical impulses and therefore affect brain-to-body-to-brain communication. Because a patient's immune system is attacking a part of the patient, the disease is called an **autoimmune** disease. Although the cause of the disease is still under study, some scientists believe that the circulatory system plays an important role, because the presence of lesions near blood vessels implicates the circulatory system.

Patients with the so-called relapsing-remitting form of multiple sclerosis experience periodic attacks interspersed with remissions. Sometimes the patients can completely recover from the neurologic damage during the remission, but in other cases at least some of the damage remains, leading to progression of the disease over time. Patients with the primary-progressive form of the disease don't have sudden attacks, but instead experience a slow, continuous worsening of their neurologic function. Individuals have widely varying symptoms, ranging from mild to severe. Some of these symptoms include:

- Balance problems and dizziness, often associated with difficulties in walking
- Bladder and/or bowel dysfunction
- Fatigue
- Memory and cognitive declines
- Numbness, especially in the extremities
- Pain, especially after remaining immobile for any length of time
- Sexual dysfunction
- Vision declines
- Seizures
- Tremor
- Speech problems, sometimes accompanied by difficult swallowing

Scientists now believe that the disease may be the result of a combination of causes, each of which makes a person more susceptible to developing multiple sclerosis. Genetics is one suspect, because the disease is more common within certain families and among certain ethnic groups. For example, if one member of a family has multiple sclerosis, that person's siblings have about a one in four chance of developing the disease, too. Geographic location and perhaps regional environmental conditions appear to have some connection, because the disease is more common in colder climates and in certain locations. Some new studies suggest that a bacterial infection may be at least partially responsible for initiating the disease. Another possible cause is exposure to toxins, including heavy metals.

No cure exists. People with MS are generally placed on a drug regimen, like an interferon, that effectively turns off the body's immune reaction and limits the continued damage to the myelin sheath. Corticosteroids are often prescribed to reduce the number and severity of attacks. Physical therapy is a common part of MS treatment, and usually includes exercises to keep the body loose and the muscles active. Alternative treatments are particularly popular among some people who have this disease, and include everything from herbal products to bee venom.

MYOCARDITIS

Myocarditis is inflammation of the myocardium, or heart muscle. A common form of this condition is **viral myocarditis**, which is a viral infection of the heart muscle walls. In extreme cases, myocarditis can lead to cardiac failure. Some patients are unaware that they have the condition. Those with indications of myocarditis typically complain of several of the following:

- Weak, abnormal heartbeat
- Chest pain above the heart
- Fever, cough, and/or nausea, and sometimes vomiting, especially in viral myocarditis
- Muscle and joint pain, especially in viral myocarditis
- Enlarged liver, especially in viral myocarditis
- Fainting spells

The causes of myocarditis include various bacterial, viral (including HIV—see separate section on AIDS earlier in this chapter), and parasitic infections. Other causes are alcohol consumption, radiation or chemical poisoning, rheumatic fever (see separate entry later in this chapter), carbon monoxide poisoning (see separate entry earlier in this chapter), heat stroke, burns, hepatitis (see separate entry earlier in this chapter), and nephritis, which is an inflammation of the kidney. For viral myocarditis specifically, enteroviruses and adenoviruses are the primary culprits.

A health professional will likely prescribe rest and relaxation until the inflammation runs its course. Sometimes a patient may also receive painkillers, like **nonsteroidal anti-inflammatory drugs (NSAIDs)**, or medications to fight the bacteria, parasite, or other cause of the condition. Treatment heightens if the patient is experiencing heart failure (see section on heart failure earlier in this chapter).

PAROXYSMAL COLD HEMOGLOBINURIA

Paroxysmal cold hemoglobinuria (PCH), also called **Donath-Landsteiner syndrome**, is a blood disorder that destroys red blood cells. Exposure to cold temperatures prompts antibodies and a cell-destroying protein to seek out and exterminate red blood cells, particularly those moving through the hands and feet, which are most affected by cold temperatures. The blood cells actually lyse (break apart) as they warm back up.

In mild cases, a patient may have no symptoms. Persons with moderate to severe attacks may experience:

- Short-lived fever, chills, and headache
- Blood in the urine
- Short-lived back pain and/or leg cramps
- Jaundice and enlarged spleen
- Symptoms of anemia (see separate section earlier in this chapter)

Some forms of syphilis, measles, mumps, influenza, and other viral and bacterial infections are often seen in association with PCH, but oftentimes the cause remains unidentified. Patients may recover on their own without medical intervention. Some, however, may become anemic and require treatment.

PERICARDITIS

Pericarditis is an infection of the pericardium, the two-layered membranous sac around the heart. Several types exist, including constrictive and fibrinous carditis. In constrictive pericarditis, the two layers of the pericardium adhere. In the fibrinous form, a thick liquid accumulates between the two pericardial layers. A fibrinous pericardium has a characteristic look, called "bread and butter," to describe the butterlike consistency of the accumulation. Patients may report one or more of the following symptoms:

- Fever
- Pain in the chest above the heart
- Difficulty breathing, sometimes with a dry cough
- Heart palpitations (see separate section on arrhythmia earlier in this chapter)
- Irritability
- Sweats, chills, and paleness of skin, especially in advanced cases
- Swelling of the chest above the heart, especially in advanced cases

Pericarditis may result from a bacterial infection, especially from such inflammations as pneumonia (lungs), meningitis (brain and spinal cord membranes), or osteomyelitis (bone and bone marrow). Pericarditis may also arise from viral infection or other disorders, including uremia (a toxic kidney condition) and hypothyroidism (inadequate thyroid secretions). A regimen of antibiotics may be prescribed, combined with bed rest and possibly painkillers for chest discomfort. For advanced cases involving chest swelling, medical professionals may recommend surgical drainage. Those with constrictive pericarditis may require surgery to remove part of the pericardium.

RAYNAUD'S DISEASE

Raynaud's disease, or Raynaud's phenomenon, is a condition brought on by cold temperatures, which trigger the arteries to contract, particularly in the extremities, and limit blood flow to these areas. The nose and ears may also be affected. Attacks can occur following even very short-term cold exposure. An attack of Raynaud's disease typically includes:

- Whitish skin, which may turn blue, then reddens as it rewarms
- Noticeably cool skin temperature
- Numbness or feeling of "pins and needles"
- In cases of extended cold exposure, possible gangrene

Emotional or physical stress seem to provoke attacks—and sometimes are sufficient to stimulate an attack even in the absence of cold temperature. In some cases, the cause may be one of several medications for cardiovascular problems and headaches, or another condition like arthritis or high blood pressure (see separate section earlier in this chapter). Often, however, the cause remains unknown. If the source of the disease can be determined, the patient will receive treatment specific to that cause. Medical care may also include medication to dilate vessels and assist blood flow. The patients are typically urged to keep themselves warm and protect all exposed areas from cold temperatures.

RH DISEASE

Also known as **hemolytic disease of the newborn** or **erythroblastosis fetalis**, **Rh disease** is the destruction of fetal red blood cells brought on by the mother's immune response.

Symptoms of Rh disease may include:

- Jaundice, resulting from bilirubin released by hemoglobin breakdown
- Extreme, generalized swelling
- Enlargement of the spleen (see section on splenomegaly later in this chapter) and liver
- Anemia (see separate section earlier in this chapter), sometimes leading to brain damage or heart failure (see separate section earlier in this chapter)

Rh disease is a severe reaction that can occur between a woman with Rh-negative blood and her Rh-positive fetus. Rh-negative women do not make Rh factor, a protein on the surface of red blood cells. Their bodies therefore view blood cells with the Rh factor as foreign and produce antibodies to eliminate the invading entities. Blood banks check for the Rh factor to prevent adverse reactions, so the only problem now arises when an Rh-negative woman carries an Rh-positive child. The Rh trait is inherited, and an Rh-positive father and Rh-negative mother can yield an Rh-positive fetus. In most cases, no problem occurs because the infant's and mother's blood supplies remain separated. In some cases, however, the mother's blood can enter the fetal blood supply, especially during labor and delivery. If this occurs, the mother's body may have time to recognize the foreign entity (a process called sensitization) and begin to produce antibodies that seek out, attack, and destroy the fetal red blood cells. Fortunately in this case, the mother's body needs some time to mount an effective immune assault, and the child is born without a problem. However, if the mother has already been exposed to the Rh-positive blood through a previous pregnancy or has been exposed to Rh-positive blood through a transfusion, her body is able to produce antibodies very rapidly and to begin destroying fetal red blood cells almost immediately, sometimes even leading to a stillbirth.

Now that the cause of Rh disease is well understood, the condition has become less and less common. Rh-negative women now receive a blood product at about seven months into the pregnancy and soon after delivery of an Rh-positive baby. This blood product, called Rh immune globulin, prevents a woman from becoming sensitized to Rh factor. Rh immune globulin actually comprises anti-Rh antibodies like those that the mother would make. When added, the antibodies quickly disperse, then latch onto and destroy any fetal blood cells in the woman's bloodstream—before her body has time to recognize that any foreign entity has invaded.

Babies who still are born with moderate to severe Rh disease typically receive transfusions with Rh-negative blood. An Rh-positive baby has no adverse reaction to Rh-negative blood—only to the antibodies made by its mother—and the mother's antibodies do not attack Rh-negative blood cells. This gives the baby an adequate blood supply until the mother's antibodies dissipate and the baby's own Rh-positive blood can recover.

Babies with mild symptoms may only require phototherapy to treat the jaundice.

RHEUMATIC FEVER

Rheumatic fever is an inflammatory disease that can result in damage to heart valves. It usually occurs in children and teens. If the valve damage ultimately affects the heart's performance, the condition is elevated to rheumatic heart disease. Patients typically complain of several mild to severe symptoms, including:

- Sudden high fever lasting 10–14 days
- Sudden joint pain, sometimes moving from joint to joint, and accompanied by swelling or reddening
- Difficulty breathing
- Chest pain
- Fatigue
- Loss of appetite
- Ashen skin
- Nosebleed
- In advanced cases, enlarged heart and/or heart murmur (see section on valvular abnormalities later in this chapter)

Strep throat, which is caused by a streptococcal infection, is a precursor to rheumatic fever. Penicillin or other antibiotics are prescribed to treat strep throat. Because individuals who have had rheumatic fever in the past are more likely to get it again, health professionals sometimes recommend periodic doses of antibiotics to prevent recurrences. If the patient has experienced valve damage, valve-replacement surgery may be recommended. An artificial valve or a valve from a pig may be used.

SHOCK

Shock is any condition that results in a sudden drop in blood flow, which in turn lowers oxygen delivery to and waste removal from tissues. This has the potential to drastically affect organ function. Generally, shock patients have several of the following:

- Ashen, cool, and clammy skin, sometimes with bluing of the lips and base of the fingernails
- Dizziness, sometimes fainting spells or unconsciousness

- Weakness
- Low blood pressure
- Sweating
- Rapid pulse
- Shallow breathing
- Anxiety and/or confusion

Shock can result from a number of conditions, including internal and external bleeding; diarrhea and burns that can lower fluids and blood volume; arrhythmias, heart attack, and heart failure that stop blood flow; the use of certain drugs that affect blood movement; and allergies that result in so-called **anaphylactic shock**. For blood or fluid loss (called **hypovolemic shock**), the standard treatment is a blood or plasma transfusion, respectively. The medical response to shock that results from flow problems generally is treatment of the source of the problem. For treatments of arrhythmias and heart failure, see the separate sections in this chapter.

SICKLE-CELL ANEMIA

Persons with **sickle-cell anemia** have slightly different hemoglobin, called Hb S, that causes red blood cells to become crescent-shaped when they give up their oxygen. These sickled cells can clump together to block blood vessels and to accumulate in and damage such organs as the liver and spleen. In addition, sickle cells die in just 10–20 days, compared to the normal red blood cells' lifespan of six months. This causes a low red blood cell count, and therefore anemia. Individuals with sickle-cell anemia report the following:

- Acute pain in areas where blood vessels become disrupted
- Hemolytic anemia (see separate section on anemia earlier in this chapter)
- Jaundice (yellowing of the skin)
- Infections that can be severe
- Leg ulcers
- Occasionally stroke

This is a genetic disorder that in the United States affects one in 500 African Americans, although many more—one in twelve African Americans—carry the genetic mutation that can lead to the disease. Although it is usually associated with persons of African descent, it can occur in persons of other ethnic backgrounds. Individuals who have inherited a mutation from only one parent generally do not have any symptoms. Those who

Malaria and Sickle-Cell Anemia

In an unusual twist, a blood-related mutation can help prevent a blood-related disease. The latter is malaria, an usually acute disease that causes chills and severe fever. Spread by *Anopheles* mosquito bites, malaria occurs when parasites enter human tissues and, in about a week, invade and destroy red blood cells. One of the best protections against malaria is the mutation for sickle-cell anemia. People with the sickle-cell trait (all those who carry one sickle hemoglobin gene) typically survive malaria untreated; those with normal hemoglobin do not.

At the same time, individuals who have inherited the mutation from both parents have sickle-cell disease. Sickle-cell disease can prove fatal, particularly if untreated. This presents a balancing act that has bolstered the presence of the sickle-cell trait in many populations. On the one hand, individuals who live in areas prone to malaria and who carry the mutation have a selective advantage in that they can survive malaria. On the other hand, individuals with mutations inherited from both parents may die from the disease.

Evolutionarily, diseases that result in early death such as untreated sickle-cell anemia usually become more and more rare because those who carry the trait do not survive to reproduce. Sickle-cell anemia does not follow this pattern, however, because the trait also confers its survival advantage. It is an odd relationship, and one that has allowed the sickle-cell mutation to continue through the generations.

have inherited the trait from both parents are described as having sickle-cell disease and have symptoms. Incidence of sickle-cell disease varies geographically, because the mutation offers some protection from **malaria** (see "Malaria and Sickle-Cell Anemia").

The only current cure for sickle-cell anemia is a bone marrow transplant, which poses some risk to the patient and is therefore usually recommended only for patients with severe symptoms. Researchers are exploring the possibility of gene therapy as a cure for the disease. Here, scientists are trying to determine ways of turning off the mutated gene while turning on another that makes fetal hemoglobin, or of swapping the defective gene for a gene from another person that makes normal hemoglobin. The work is ongoing.

To treat symptoms, medical professionals may prescribe antibiotics to stave off infections, and painkillers. Some patients may receive **hydroxyurea**, which was formerly used only to fight tumors. This drug prevents the sickling of blood cells and the associated painful symptoms. In addition, blood transfusions may be used to boost the red blood cell count, and devices called incentive spirometers to deter lung complications.

Current studies are investigating agents that increase the production of fetal hemoglobin in a patient's bloodstream. Usually, the production of this type of hemoglobin stops at birth, but some patients with sickle-cell disease continue to produce fetal hemoglobin after birth. Comparisons indicate that they have far milder symptoms that those patients who do not make fetal hemoglobin. Scientists believe that hydroxyurea works by stimulating the production of fetal hemoglobin, and are now looking at other chemicals, like **butyrate**, with a similar function.

Researchers are also considering the possibility of using gene therapy to prevent sickle-cell disease. In research published in the December 14, 2001, issue of *Science*, a group from Harvard Medical School and the Massachusetts Institute of Technology was able to prevent the disease in mice. They accomplished the feat by indirectly adding a modified gene to stem cells in the bone marrow. Within a year of adding the gene, they reported that all of the mice were symptom-free.

SPHEROCYTOSIS

Spherocytosis, also known as **congenital sperocytic anemia**, affects the membranes of red blood cells and changes them from their normal disk shape to spheres. The spleen detects the abnormal cells and destroys them. Common symptoms include:

- Anemia (see separate section earlier in this chapter)
- Splenomegaly
- Occasionally stomach ache and/or loss of appetite
- Weakness
- Blood in the urine
- Fever and/or vomiting
- Jaundice or ashen skin, especially in infants

This is a genetic disorder, caused by a defect in a protein that comprises the red blood cell's surface. Patients may follow the general treatments listed in the separate section on anemia. Physicians sometimes recommend a splenectomy to remove the spleen and alleviate some of the premature destruction of the abnormal red blood cells.

SPLENOMEGALY

Splenomegaly is an enlargement of the spleen—sometimes to ten times or more of its normal size, which is about 7 ounces (200 g). This spongy organ that lies below the stomach has two primary functions: assisting in the removal of foreign materials and aging red blood cells from the circulation; and storing red blood cells and platelets. Sometimes, this condition may become so extreme that the spleen ruptures and begins to release blood into the abdomen. Beyond an enlargement of the spleen, symptoms vary widely depending on the cause of the enlargement. A ruptured spleen carries its own set of symptoms, including abdominal pain on the left side, dizziness, and difficulty breathing.

The spleen may enlarge for a number of reasons, but usually it results from thalassemia (see section on anemia) and other hemolytic anemias, as well as other conditions that destroy blood cells. Various viral infections, cirrhosis of the liver (see separate section earlier in this chapter), cystic fibrosis, malaria (see "Malaria and Sickle-Cell Anemia") and cancers like Hodgkin's disease and leukemia (see separate section earlier in this chapter) are common precursors to splenomegaly.

Treatment varies widely, but typically is targeted at the cause of the condition rather than the condition itself. A ruptured spleen is a serious problem that generally results in a splenectomy, or surgical removal of the organ.

STROKE

Often called a "brain attack," a stroke is a sudden blockage of blood flow to a part of the brain. It can result in brain cell death, and loss of functions that are controlled by those cells. Common poststroke problems include loss of speech or movement on one side of the body. Stroke is sometimes described as cerebrovascular disease (CVD). More than 500,000 strokes are diagnosed in the United States annually. Many stroke symptoms are characterized by their sudden onset. Patients typically complain of one or more of the following:

- Sudden numbness or weakness of one side of the body, typically manifesting in the face, arm, or leg
- Sudden loss of communication abilities, including the ability to speak or to understand speech
- Sudden loss or impairment of vision
- Sudden, unexplained headache that is severe in nature
- Sudden dizziness or loss of balance

Some people also have "ministrokes." These are called **transient ischemic attacks (TIAs)**, and they are temporary strokes. In TIAs, the blood clot dissolves before it has caused a long-enough stoppage of blood flow to bring about lasting effects. TIAs may or may not precede major strokes. Despite their temporary nature, TIAs should be treated as a medical emergency, and the patient should seek medical help immediately.

A stroke may occur in one of two ways. In the more common of the two, called cerebral infarction or **ischemic stroke**, an artery's blood flow is disrupted and the affected part of the brain begins to die. Most of the time, the disruption is due to a blood clot that either forms in the artery itself or breaks loose from another part of the body, travels to the brain, and becomes lodged in an artery there. The type of stroke usually involves a blockage in the carotid or basilar artery. Stroke may also result from bleeding in the brain that results when an artery suddenly bursts at a weak spot, called an **aneurysm**. This condition, known as a **hemorrhagic stroke**, may result in bleeding inside the brain or in the space between the brain and the skull.

Treatment for ischemic stroke is very different from that for hemorrhagic stroke. All persons presenting with stroke symptoms should report immediately to an emergency room at a hospital. Hospital staff will perform brain scans to verify the diagnosis of a stroke and to distinguish which of the two types of stroke has occurred. **Computed tomography (CT) scans** are commonly used tools for making these determinations. In addition, medical providers may use other tools, like **magnetic resonance imaging** or **transcranial doppler**, to view blood flow and to locate sites of blockage (see color insert).

Time is of the essence for the treatment of stroke. In the case of ischemic stroke, a group of drugs called **tissue plasminogen activators (tPAs)** is used to prevent the progression of brain damage and even reverse it. These so-called clot-busters are so effective at dissolving clots and restoring blood flow that many patients make a full recovery—with no lasting effects—within a few months. The drugs do carry some risk of causing uncontrolled bleeding, a risk that heightens over time and becomes too great after three hours from the onset of symptoms. Before their use, it is essential to verify the type of stroke, because they can encourage bleeding in patients with hemorrhagic stroke. Besides tPAs, physicians may recommend that a patient undergo surgery to have the blocked artery reopened. This procedure, known as carotid endarterectomy, removes plaque from the carotid artery and improves blood flow.

The treatment for hemorrhagic stroke may include drugs to reduce swelling (recall that fluids may accumulate in brain tissues or between the brain and skull), or surgery either to pinch off the bleeding artery or to drain away excess blood that is putting pressure on the brain. New research indicates that other drugs may help to restore brain function following a stroke. In the September 8, 2001, issue of *Lancet*, researchers from the Neurology Clinic in Bad Aibling, Germany, reported that the drug levopoda, which is used to treat Parkinson's disease, appeared to aid recovery in half of the patients tested. They gave the drug to twenty-two wheelchair-bound stroke patients, and within six weeks, eleven were walking—nearly twice the rate seen in patients who did not receive the drug. Almost simultaneously, U.S.

researchers reported in the September 2001 issue of *Stroke* that the stimulant dextroamphetamine promoted recovery of speech in stroke patients.

Other options for stroke patients include the use of catheters to open blocked arteries in ischemic stroke, or to close bleeding arteries in hemorrhagic stroke. Blood thinners are commonly prescribed to discourage the formation of new blood clots.

VALVULAR ABNORMALITIES

Valvular abnormalities may be either a **valvular stenosis**, or narrowing of the heart valve, or an **incompetent valve**, which closes incompletely. One of the most common valvular abnormalities is a **mitral stenosis**, in which the **mitral valve** may be a half to a tenth of the normal size. Aortic valve stenoses are also common. Such a narrowing puts extra demands on the heart, which must pump more forcefully to ensure adequate blood flow. In addition to damage and possibly heart failure caused by the additional work, a stenosis may cause the heart to enlarge or may lead to leakage of the blood backwards through the valve. This leakage is called **valvular regurgitation**. People who have valvular abnormalities often have **heart murmurs**, sounds produced by the vibration of blood as it traverses the faulty valve. Heart murmurs, however, do not always signify a valve problem. In fact, heart murmurs are quite common in children. These murmurs, called **benign systolic murmurs**, result from blood flowing irregularly from the ventricle and pose no health threat. Other symptoms, if any are present, include:

- Shortness of breath, particularly while lying down, and coughing
- Weakness and/or fatigue
- Chest pain
- Swelling of the extremities
- Heart palpitations (see separate section on arrhythmia earlier in this chapter)
- Dizziness, sometimes fainting and
- Cyanosis, or bluish skin

Valvular abnormalities may be genetic in origin but also may result from such conditions as rheumatic fever (see separate section earlier in this chapter), atherosclerosis (see separate section earlier in this chapter), or an accumulation of calcium that can bog down valves.

The treatment for valvular abnormalities typically involves anticoagulant drugs to assist blood flow through the valve, and other medications to treat arrhythmias, edema (swelling), or other symptoms. Severe conditions may also call for valve-replacement or -repair surgery. Valve-repair surgery may entail widening the valve by forcing it open with a balloon catheter (see the section on heart attack earlier in this chapter) or by cutting away some of the valve if it is thickened.

VON WILLEBRAND'S DISEASE

People with **von Willebrand's disease** are deficient in the blood-clotting factors, known as factor VIII and von Willebrand's factor. This leads to uncontrolled bleeding. Type 3 is the most severe form of the disease and can lead to more severe bleeding problems. This disease, which is sometimes mistaken for hemophilia (see separate section earlier in this chapter), is often asymptomatic. When complaints exist, they may include:

- Bruising with little cause
- Nosebleed
- Excessive and spontaneous bleeding from mucus membranes or in the gastrointestinal tract
- Excessive bleeding following injury or other trauma
- In women, excessive menstrual bleeding
- Anemia

Von Willebrand's disease is mostly genetic in origin, although some individuals apparently can acquire the disease after experiencing certain other immune disorders. Von Willebrand's disease typically appears early in life and eases as the person ages.

During bleeding episodes or before surgery, medical professionals may prescribe a hormone that stimulates the body to produce the deficient factors, or give the patient supplements of the factor. Other medications may also be used.

WEST NILE FEVER

West Nile fever, new to the United States in 1999 but long known to Africa, Asia, and Europe, is a viral infection that is spread by birds and mosquitoes (see photo). It can result in an inflammation of the brain. It is one of many so-called blood-borne viruses that are transferred from an infected organism to the blood of another organism. People with West Nile fever often don't have symptoms or have very mild symptoms. Others, especially infants and elderly people, may be seriously affected and can die. Symptoms of West Nile fever may include:

- Headache, ranging from mild to severe

- Fever, ranging from mild to high

- Rash and/or eye irritation

- Swollen glands

- Muscle weakness or stiffness

- Coma, possibly death, in serious cases

The Asian tiger mosquito, vector of the West Nile virus, sits on a human finger. Humans who contract the virus may develop West Nile fever, which is sometimes fatal. © Centers for Disease Control and Prevention.

Rarely, the disease may lead to encephalitis or meningitis, which are inflammations of the brain or of the membranes of the brain and spinal cord, respectively.

West Nile fever arises from infection with a virus. Humans are infected following a bite from an infected mosquito (*Culex* species). The mosquitoes themselves become infected after they bite a bird that carries the virus. Rarely, the West Nile fever virus is spread through blood transfusion or from mother to infant through breastfeeding. Most people recover from the disease on their own. Infants, elderly persons, and individuals with compromised immune symptoms, however, may require treatment for their individual symptoms.

Lifestyle Choices for Cardiovascular Health

Study after study concludes the same thing: A healthy lifestyle promotes cardiovascular health. For years, physicians have been touting the benefits of diet, exercise, and lowered anxiety, and scientific research is continuing to confirm the veracity of their lectures.

BENEFITS OF DIET

An overwhelming consensus of medical professionals is that a diet rich in vegetables and fruits, but low in fats, is important to cardiovascular health.

The National High Blood Pressure Education Program issued a major report published in the October 16, 2002, edition of the *Journal of the American Medical Association*, which updated its prevention guidelines for high blood pressure. These guidelines resulted from a large clinical trial known as Dietary Approaches to Stop Hypertension, or DASH. According to the guidelines, fruits, vegetables, and lowfat dairy products should be incorporated into the diet, but total fat, and especially saturated fat, should be limited. In addition, daily alcohol consumption for men should be limited to 1 ounce of ethanol, which equates to two 12-ounce cans of beer, 10 ounces of wine, or 2 ounces of 100-proof whiskey. Women should slash those alcohol totals in half.

The National High Blood Pressure Education Program went on to recommend that individuals with high blood pressure cut their potassium intake to no more than 3,500 milligrams per day, and sodium intake to a maximum of 2,400 milligrams per day. In results presented at the 2000 annual meeting of the American Society of Hypertension, researchers demonstrated that individuals who followed the diet recommendations and also cut back their use of salt were able to trim blood pressure by 8.9 mmHg on the systolic side, and 4.5 mmHg on the diastolic—regardless of whether they had high blood pressure at the outset.

According to Dr. Paul Whelton, co-chair of the working group that developed these recommendations, "Epidemiological data suggest that if we could lower the average systolic blood pressure among Americans by 5 mmHg, we'd see a 14 percent drop in deaths from stroke, a 9 percent drop in heart disease deaths, and a 7 percent drop in overall mortality."

The National Cholesterol Education Program developed a similar set of guidelines for individuals who are trying to control or lower their cholesterol levels. It recommends that a maximum of 7 percent of the daily caloric intake come from saturated fat, and that cholesterol intake be limited to 200 milligrams per day. It suggests the addition of soluble fiber, which is found in some fruits, vegetables, and cereal grains, to lower the level of low-density lipoprotein, also known as LDL or "bad" cholesterol. Food products containing plant stanols or plant sterols can have the same effect.

In addition to these general recommendations for cardiovascular health, a multitude of studies have been conducted to determine the positive and negative aspects of certain foods. Some of these include the following.

Coffee

Several recent studies have indicated that coffee can raise cholesterol levels. To test whether reductions in coffee consumption could lower those levels, researchers at Ulleval University in Norway collected cholesterol levels in 191 people, then asked them to follow one of three regimens over a six-week period. The study, published in 2001, showed that those who continued to drink four cups a day saw no change in their cholesterol level, nor did those who cut back to three or less cups of filtered coffee. Those who quit the habit completely, however, showed blood cholesterol drops of about 5 percent, a well as a drop of 12 percent in homocysteine level, which is an amino acid associated with a heightened risk for cardiovascular problems.

Bananas

Because of its high potassium content, bananas can help lower blood pressure, particularly in those with hypertension. According to a 1998 study in *Hypertension*, potassium lowers systolic and diastolic blood pressure by about two points even in healthy individuals. Other studies indicated about twice the effect in people with hypertension. The 1998 study noted that potassium can be in the form of bananas or any other fruit or vegetable high in the mineral, or by way of potassium supplements.

Tea

While coffee poses risks to the cardiovascular system, research indicated that tea has benefits. A researcher at the Boston University School of Medicine set out to determine why black-tea drinkers were less likely to develop heart disease, and found that black tea dilates blood vessels. The researcher, Joseph Vita, found that consumption of two 8-ounce cups of tea was enough to improve the blood flow in patients who had atherosclerosis.

Wine

"Everything in moderation" is the saying, and it applies to wine, too. Numerous studies have linked wine to cardiovascular health benefits. In an attempt to understand these positive attributes, researchers at Queen Mary, University of London, looked at blood vessels. Following lab experiments, they reported in 2002 that wine consumption caused a decrease in a chemical that made the vessels constrict, but only if the wine was red. Whites and rosés had no effect.

Grape Juice

According to research presented at the 2003 annual meeting of the Federation for American Societies of Experimental Biology, Concord grape juice was effective at lowering blood pressure. The study of 80 men aged 45 to 70 showed that those with hypertension were able to lower their systolic and diastolic blood pressures by about five points after drinking 12 ounces of the juice every day for three months. At least two other studies of grape juice also showed that its consumption enhanced the ability of arteries to expand in response to blood-flow increases.

Diabetes

Type 2 diabetes is one of several cardiovascular-related diseases that can sometimes be prevented through proper diet and exercise. According to results announced in the September 2001 issue of *The New England Journal of Medicine*, the combination of a lowfat diet and daily exercise was enough to at least delay the onset of the disease among those who are at high risk of developing diabetes. The study also provided evidence that the drug metformin would delay onset, but it was not as effective as exercise and diet. Another small but intriging study in the May 2002 issue of the *American Journal of Clinical Nutrition* promoted whole grains as an avenue to a diabetes-free life. According to the research, a whole-grain diet helped moderate blood sugar and keep adult-onset diabetes at bay in overweight individuals.

Vegetarians

Although vegetarianism is often considered the epitome of a healthy lifestyle, scientists warn that some vegetarians may be setting themselves up for heart disease. In a study published in the February 2002 issue of the *Journal of Nutrition*, a research group noted that many vegetarians are lacking in certain essential nutrients, including vitamins B_6 and B_{12}. These nutrients are present in meat products, but not fruits and vegetables. The vitamin B deficiencies can in turn lead to an overaccumulation of the amino acid homocysteine in the blood, which has been shown to increase the risk for heart disease. The B vitamins are needed for metabolizing homocysteine. The bottom line is that vegetarians need to eat foods fortified with the vitamins or take vitamin supplements to counter the risk.

BENEFITS OF EXERCISE

Like a healthy diet, exercise is an important part of a cardiovascular-healthy lifestyle, and numerous studies have verified its significance.

In September 2002, the National Academies' Institute of Medicine issued new recommendations, including at least 60 minutes of moderately intense exercise every day. The report stated that the 60 minutes could be spread over the day, so a person might accumulate the hour by taking a quick walk to and from a corner store instead of driving, climb the stairs a few times instead of riding in an elevator, and take a bike ride after work instead of watching TV from the couch.

Exercise isn't just for the healthy. In the March 4, 2003, issue of *Circulation*, the American Heart Association recommended exercise for a broad range of people, even heart-failure patients who are awaiting a heart transplant. The association said that exercise can promote proper functioning in blood vessels, improve the delivery of oxygen to muscle tissues, and decrease hormones that bring on symptoms of heart failure. While it stressed that heart-failure patients should work closely with their health-care professionals in developing

Compared to an out-of-shape person, an athlete's heart beats slower and stronger. An extremely fit individual's heart may pump up to seven times more blood per beat. © Photodisc.

an exercise plan, the association suggested that patients would benefit from 20–30 minutes of exercise on three to five days each week.

A study at Duke University Medical Center, Johns Hopkins, Pennington Biomedical Research Center, and the Center for Health Research found that patients were able to implement successfully the DASH diet and an exercise plan consecutively, and avert high blood pressure. The study was published in the April 23, 2003, issue of the *Journal of the American Medical Association*.

TOBACCO AND ILLICIT DRUGS

Because cigarettes and so many illicit drugs are addictive and have deleterious effects on health, medical professionals continue to urge their patients and the general public to avoid them. Unfortunately, first-time users are usually teens and young adults who either fail to see the long-term consequences of their actions or ignore them, and become habitual users. Nonetheless, thousands of people have broken the chains of tobacco and drugs, and are now back on the path to healthy living.

Numerous studies have bolstered the medical community's outcries against cigarettes and illicit drugs, and have tied the two to cardiovascular problems. A few of those studies are mentioned here.

Smoking

Smoking is widely recognized as a risk factor for any number of cardiovascular disorders, and new research continues to add evidence of its dangers. One such study appeared in the 1997 issue of *Circulation*. That research, conducted by scientists at the Hippokration Hospital and the University of Athens, Greece, provided evidence that cigarette smoke almost immediately affected the aorta by making it less elastic within a single minute following the first inhalation. The effect persisted for at least 20 minutes after the individual took the last puff. During this period of dampened aortic flexibility, the heart responds by pumping harder, which could have dire consequences especially for heart-disease patients.

In addition, a study reported in the July 2001 issue of *Nature Medicine* suggests that nicotine, one of the chemicals in cigarette smoke, appears to trigger the growth of new blood vessels, a process called angiogenesis—at least in mice. While the growth spurt may be good news if a patient needs new blood vessels to help build heart muscle, angiogenesis is also required for the growth and spread of tumors, indicating that it also may heighten an individual's risk for cancer.

Marijuana

Numerous studies have pointed to connections between marijuana use and cardiovascular problems, including increases in blood pressure and decreases in oxygen to cardiac muscle. Adding to those findings, a large study of nearly 4,000 people showed that marijuana use is associated with an increased risk for heart attack. The study, published in the June 12, 2001, issue of *Circulation*, showed that individuals are at least three times more likely to have a heart attack if they have smoked marijuana in the past hour.

Cocaine

According to several recent cocaine studies, the use of this drug can have extreme cardiovascular consequences. A large study published in the May 31, 1999, issue of *Circulation* found that the risk of having a heart attack is twenty-four times higher if an individual has used cocaine in the past hour. That heightened risk exists regardless of whether the individual had any prior heart-disease symptoms. The study included data collected from nearly 4,000 people.

A few months later, research presented at the November 9, 1999, Scientific Sessions of the American Heart Association reported that cocaine users also faced a risk for an aneurysm that was three times higher than the risk nonusers faced. Those who used cocaine the most frequently were at the highest risk.

Acronyms

ACE	angiotensin-converting enzyme	**ECG (also EKG)**	electrocardiograph
AIDS	acquired immunodeficiency syndrome	**FSH**	follicle-stimulating hormone
AV node	atrioventricular node	**Hb S**	hemoglobin S
AV valve	atrioventricular valve	**HCO_3^-**	bicarbonate ion
AZT	zidovudine	**HDL**	high-density lipoprotein
bpm	beats per minute		
Ca^{2+}	calcium	**HHT**	hereditary hemorrhagic telangiectasia
Cl^-	chloride	**HIV**	human immunodeficiency virus
cm	centimeter		
CO_2	carbon dioxide	**HPO_4^{2-} and $H_2PO_4^-$**	phosphate
CPR	cardiopulmonary resuscitation	**Ig**	immunoglobulin
		IgA	immunoglobulin A
CT scan	computed tomography scan	**IgD**	immunoglobulin D

IgE	immunoglobulin E	**NaCl**	sodium chloride (table salt)
IgG	immunoglobulin G		
IgM	immunoglobulin M	**NaHCO$_3$**	sodium bicarbonate
ITP	immune thrombocytopenic purpura	**NIDDK**	National Institute of Diabetes and Digestive and Kidney Diseases
K$^+$	potassium		
l	liter	**nm**	nanometer
LDL	low-density lipoprotein	**NSAIDs**	nonsteroidal antiinflammatory drugs
LH	luteinizing hormone	**O$_2$**	oxygen
m	meter	**PCH**	paroxysmal cold hemoglobinuria
Mg^{2+}	magnesium		
mL	milliliter	**PS**	lipid phosphatidylserine
mm	millimeter	**SA node**	sinoatrial node
mmHg	millimeters of mercury	**SO$_4$$^{2-}$**	sulfate
MRI	magnetic resonance imaging	**TIA**	transient ischemic attack
MS	multiple sclerosis	**tPA**	tissue plasminogen activator
Na$^+$	sodium	**μm**	micrometer

Glossary

ACE inhibitors *See* angiotensin-converting enzyme inhibitors.

acquired immunodeficiency syndrome (AIDS) A medical condition in which immune function is severely depressed. As a result, the body has little defense against disease, including everyday infections.

adrenal gland A gland located on top of the kidney. It releases the hormone adrenaline.

adrenaline Also called epinephrine. Used in the body's fight-or-flight response, it is responsible for the quickened heartbeat, rapid breathing, and other responses that accompany stress.

agglutination The clumping of blood. This can occur if a patient with a certain blood type is given blood of another type.

AIDS See acquired immunodeficiency syndrome.

albumin The most abundant plasma protein. It makes up 55 percent of the total protein content of plasma. It is involved in maintaining blood volume and water concentration.

alveoli Tiny air sacs in the lungs. They exchange oxygen and carbon dioxide between the lungs and the blood.

amino acids The building blocks of proteins.

anaphylactic shock Shock brought on by an allergy.

anemia A medical condition in which patients have too few red blood cells or too little hemoglobin in their circulating blood, leading to insufficient oxygen delivery to body tissues.

aneurysm Bleeding in the brain that results when an artery suddenly bursts at a weak spot. It results in hemorrhagic stroke.

angina pectoris Temporary chest pain or pressure that radiates from the heart to the shoulder and left arm, or sometimes from the heart to the abdomen. Angina occurs when the heart muscles receive insufficient oxygen, typically from inadequate blood flow.

angiogenesis The growth of new blood vessels.

angioplasty A medical procedure in which a catheter is threaded through the blood vessels and to a narrowed artery. The end of this catheter may contain some type of device, such as a balloon, to widen the vessel.

angiotensin-converting enzyme (ACE) inhibitors Often used to treat heart attacks and other heart conditions, this group of drugs widens blood vessels, and therefore reduces the demands on the heart.

anions Negatively charged particles.

annulus fibrosus A ring of fibrous connective tissue that serves as an anchor for the heart muscle and as an almost-continuous electrical barrier between the atria and ventricles.

antibody Proteins that attack antigens.

antigens Invading organisms and materials that enter the human body. The body may mount a defense with antibodies.

aorta The largest artery in the human body, it is the one leaving the heart with newly oxygenated blood.

aortic aneurysm See aortic dissection.

aortic arch A large, rounded section of the aorta that occurs above the heart, just after the aorta leaves the right ventricle.

aortic dissection Also known as an aortic aneurysm. It is a medical condition brought on by a tear in the lining of the aorta. It results in bleeding into the aorta wall, sometimes leading to a

blood-flow stoppage, heart failure, or rupture of the aorta.

aortic semilunar valve The three-cusped heart valve located between the left ventricle and aorta.

aplastic anemia An uncommon form of anemia in which the individual's bone marrow is unable to produce blood cells.

arrhythmia Heart rhythms that deviate from the normal "lubb-dupp" pattern and pace of the heart. Arrhythmias include fast beating, or tachycardia; slow beating, or bradycardia; lack of beating, or asystole; and other abnormal patterns.

arterial anastomosis An alternative site used by organs such as the brain to deliver blood if a supplying artery becomes blocked.

arterial baroreceptor reflex The mechanism that provides oversight and maintenance of the blood flow by responding to slight changes in blood pressure.

arterial system The portion of the circulatory system that delivers oxygen-rich blood to the body tissues.

arteries Larger blood vessels that deliver oxygen-rich blood to the body tissues.

arterioles Smaller blood vessels that deliver oxygen-rich blood to the body tissues.

arteriovenous anastomoses Blood vessels that directly connect arterioles to venules. Commonly, blood travels from arterioles to capillaries to venules. Arteriovenous anastomoses are typically found in only a few tissues.

asystole Arrhythmia in which the heartbeat stops.

atherosclerosis Also known as hardening of the arteries. It is a narrowing of arterial walls caused by deposits, collectively called plaque, that create rough, irregular surfaces prone to blood clots.

atria Plural of atrium.

atrial fibrillation Rapid beating of the atria.

atrioventricular node See AV node.

atrium In the human heart, it is one of the heart's two upper chambers. The plural form is atria.

atropine A drug that may be prescribed to activate the heart muscle. It may be used to treat bradycardia or asystole.

autoimmune A term used to describe an immune response to the patient's own body. An autoimmune disease is therefore one that attacks part of the patient's body.

autonomic nervous system The part of the nervous system that controls involuntary actions and rules the variations of the heart rate.

AV node Also known as the atrioventricular node. A small group of cells and connective tissue located at the bottom of the septum. The SA node sends its electrical impulse to the AV node, which is the only conducting path through the annulus fibrosus and indirectly to the ventricles. See bundle of His and Purkinje fibers.

AV valve Also known as the tricuspid atrioventricular valve. It is a three-cusped heart valve located between the right atrium and right ventricle.

axillary vein A vein that collects blood from the basilic and brachial veins and delivers it into the chest.

AZT (zidovudine) A drug used to treat infections associated with the human immunodeficiency virus that causes AIDS.

baroreceptors Pressure detectors located in the major arteries. Part of the arterial baroreceptor reflex, they sense a dip or spike in blood pressure.

basilar artery A blood vessel that arises from the vertebral arteries and joins with other cerebral arteries to form the circle of Willis.

basilic vein A vein that collects blood from veins in the hand and delivers it into the axillary vein.

basophil A type of granulocyte that appears to be active in the inflammatory process.

Bayliss myogenic response The mechanism by which smooth muscle cells impart muscle tone to the blood vessels.

B cells Also known as B lymphocytes. They are one of two main types of lymphocyte, and participate in the body's immune response.

benign systolic murmur A heart murmur that commonly occurs in children.

beta blockers Often used to treat heart attacks and other heart conditions, this group of drugs reduces the demands on the heart. Beta blockers inhibit adrenaline and other substances that trigger heart muscle action, causing the heart to beat slower and less forcefully.

bicuspid mitral valve See mitral valve.

bile acid sequestrants A group of drugs sometimes used to treat high cholesterol levels.

biliary cirrhosis A form of cirrhosis that is characterized by inflammation of bile ducts.

blastocyst A ball of cells that forms from a fertilized egg in early embryonic development.

blood The fluid that contains the plasma, blood cells, and proteins, and carries oxygen, carbon dioxide, nutrients, waste products, and other molecules throughout the body.

blood cells Cells contained in the plasma of the blood. See red blood cells and white blood cells.

blood clot A semisolid clump of coagulated blood.

blood pressure The force of the blood against the walls of the blood vessels.

blood-sugar level The amount of glucose in the blood.

blood type A form of blood, determined by the presence or absence of chemical molecules on red blood cells. A person may have type A, B, O, or AB blood.

blood vessels Also known as the vasculature. These are the tubes of the circulatory system that transport the blood throughout the body.

B lymphocytes See B cells.

Bohr shift The effect of pH on the oxygen affinity of hemoglobin.

bone marrow Spongy, connective tissue inside many bones.

Bowman's capsule The bulb surrounding the glomeruli. It provides an efficient transfer site for water and waste products to move from the blood to the urinary system.

brachial artery The blood vessel in the upper arm that accepts blood from the subclavian artery by way of the axillary artery, and travels down the arm to supply the ulnar, radial, and other arteries of the forearm.

brachial vein The blood vessel that collects blood from the ulnar vein and empties into the axillary vein.

brachiocephalic artery Also known as the innominate artery. This short blood vessel arises from the aortic arch, and branches into the right common carotid artery and the right subclavian artery.

brachiocephalic veins Also known as the innominate veins. This pair of veins arises from the convergence of the internal jugular and subclavian veins and flows into the superior vena cava.

bradycardia A slow-beating arrhythmia.

brain attack See stroke.

bronchial vein One of two main blood vessels that collect newly oxygenated blood from the bronchi and a portion of the lungs, and deliver it through one or more smaller veins to the superior vena cava.

bundle of His A thick conductive tract located in the heart that transmits the electrical signal from the AV node to the Purkinje fibers in the base of the ventricle wall.

butyrate A chemical under consideration for the treatment of sickle cell disease.

calcium channel blocker A medication used to treat high blood pressure and other cardiovascular conditions. It inhibits calcium, which blood vessels require for contraction. With lower levels of calcium, the blood vessel walls are more relaxed, allowing blood to flow more freely.

capillary The tiniest blood vessels. They are the sites of exchange: At body tissues, blood in the capillaries delivers oxygen and nutrients, and picks up carbon dioxide and waste products; and at the lungs, blood in the capillaries drops off carbon dioxide and picks up oxygen.

carbon monoxide poisoning A medical condition arising when a person is exposed to carbon monoxide gas. Prolonged exposure can be fatal.

cardiac muscle Heart muscle. It has characteristics of both smooth muscle and striated muscle.

cardiac output See stroke volume.

cardiomyopathy A disease of the heart muscle that makes it weak and unable to function effectively.

cardiopulmonary resuscitation (CPR) A well-known and often-taught first-aid measure to continue respiration and blood circulation in a patient whose heart has stopped beating.

carotid pulse The palpable pulse issuing from either side of the front of the neck below the jaw line.

catheter A long, narrow tube typically used in angioplasty.

cation A positively charged particle.

cavernous angiomas Abnormal blood vessels that have a collection of tiny, leaky pouches filled with blood.

celiac artery The blood vessel that arises from the abdominal aorta and distributes blood to the left gastric, common hepatic, and splenic arteries.

cephalic vein The long blood vessel that collects some of the blood from the hand and delivers it to the axillary vein.

cerebral infarction See ischemic stroke.

cerebrovascular disease See stroke.

cholesterol A fatlike substance that occurs naturally in the body. Two types exist: high-density lipoprotein (HDL) and low-density lipoprotein (LDL).

chordae tendineae Tiny tendinous cords located at each of the heart valves. They attach to nearby muscles and prevent blood backflow through the valves.

Churg-Strauss syndrome An inflammation of the blood vessels that can lead to organ failure and sometimes death.

circle of Willis A vascular structure that supplies blood to the brain. It arises from the basilar, internal carotid, and other arteries.

circulatory system The heart, blood vessels, and blood.

cirrhosis of the liver A chronic degenerative condition resulting in substantial liver damage.

clofibrate A type of drug sometimes used to treat high cholesterol levels.

coarctation of the aorta A congenital heart disorder that involves a narrowing of the aorta.

colic arteries Divided into right, left, and middle colic arteries, all of which branch from either the inferior or superior mesenteric arteries, and feed the colon.

collagen A protein of blood vessels that provides a stiffness to vessels, in contrast to the protein elastin that permits the vessels' elasticity.

collateral arteries Vessels in the heart that serve as a backup blood delivery system when a coronary artery is blocked.

colostrum The fluid that is secreted by a female's mammary glands right after childbirth and before the glands begin to produce milk.

common carotid arteries One of two major blood vessels that supply the head. The left carotid splits directly from the aortic arch between the bases of the two coronary arteries. The right carotid indirectly branches from the aorta via the brachiocephalic artery.

complement The collective term for a variety of beta globulins. See globulins.

computed tomography (CT) scan Commonly used tool for determining the nature of a stroke.

concentration gradient The change in solute concentration from one location to another. Unless restricted, solutes will move from a site of higher solute concentration to one of lower solute concentration, leading to an equilibrium between the two sites.

congenital afibrinogenemia A medical condition in which patients either lack the plasma protein called fibrinogen, which is one of the proteins that participates in blood clotting, or have defective plasma fibrinogen that is unable to function in blood clotting. The blood in persons with this disorder cannot coagulate.

congenital sperocytic anemia See spherocytosis.

congestive heart failure Heart failure accompanied by fluid in the lungs and extremities.

continuous capillaries The least permeable type of capillary. They are found in the skin, muscles, lungs, and central nervous system.

coronary angiography A medical procedure in which a catheter is used to deliver a dye into the heart's arteries so that a cardiologist can have a clear view of blood flow in the arteries and can discern whether any arteries are blocked or narrowed, and are hampering blood flow.

coronary arteries Arising from the base of the aorta, these are the two major arteries that feed the heart muscle. The right coronary artery remains a single, large vessel, but the left coronary artery almost immediately splits into transverse and descending branches.

coronary artery bypass grafting See coronary artery bypass surgery.

coronary artery bypass surgery A medical procedure that reroutes blood around a blocked or severely narrowed coronary artery. It involves clipping out the affected portion of artery and replacing it with a piece of blood vessel taken from elsewhere in the body.

coronary atherectomy A medical procedure in which a catheter is threaded through the blood vessels and to a narrowed vessel where a device is used to actually cut through plaque and reopen the lumen.

coronary circulation The circulatory system of the heart.

coronary heart disease An umbrella term used to describe many heart-related conditions, all of which involve narrowed coronary arteries that affect the blood flow to the heart.

corpora cavernosa Two columns of erectile tissue, located on either side of the penis.

corticosteroids A family of drugs used in various conditions, including multiple sclerosis.

CPR See cardiopulmonary resuscitation.

crista dividens A routing mechanism in fetal circulation that delivers blood from the right atrium to the left atrium. Adults lack the crista dividens, at the edge of the foramen ovale, and blood flows from right atrium to right ventricle.

CT scan See computed tomography scan.

cyanosis Medical term for bluish skin.

D antigen See Rh factor.

defibrillator A device that detects arrhythmias and delivers a mild electrical jolt that is just powerful enough to put the heart back on a regular pace. The term also refers to a more powerful device used by emergency personnel to revive patients whose hearts have stopped.

diabetes A medical condition in which the body either produces too little insulin or does not use it properly. This causes glucose to accumulate in the blood, then leave the body as a waste product without having nourished the cells.

diabetes mellitus Also known as adult-onset diabetes or type 2 diabetes. In type 2 diabetes, the cellular response to insulin is impaired, and the cells do not take up glucose as they should.

diastole The heart's resting period.

diffusion The passive flow of molecules from one location to another.

digitalis A medication used to boost the force of the heart's contractions; it increases blood pressure and promotes adequate oxygenation of body tissues. This may be used to treat heart failure and other conditions.

digoxin A drug used to retard electrical conduction in the heart and slow its beat. It may be prescribed to treat arrhythmias.

dilated cardiomyopathy Also known as congestive cardiomyopathy. It leads to an enlargement of one or more heart chambers.

discontinuous capillary Also known as sinusoidal capillary. They have large apertures that allow the transport of large proteins and red blood cells. They are the most water- and solute-permeable of the three types of capillaries. See fenestrated capillary and continuous capillary.

diuretics Medications that trigger the kidneys to remove more water and sodium from the plasma, thereby reducing swelling in the extremities and also lowering the total blood volume. The latter serves to reduce the heart's workload. They may be used to treat heart failure and other conditions.

Donath-Landsteiner syndrome See paroxysmal cold hemoglobinuria.

ductus arteriosus A shunt found in the fetal circulation that routes blood from the still-nonfunctioning lungs, and to the descending aorta.

ductus venosus A shunt found in the fetal circulation that allows about half of the blood from the umbilical vein to bypass the liver and directly enter the inferior vena cava.

ECG See electrocardiograph and electrocardiogram.

echocardiography A procedure used especially in heart patients to take an ultrasound of the moving heart.

ectopic pacemaker Any set of heart cells that take over the typical pacemaker function of the heart. Normally, the sinoatrial node serves as the pacemaker.

edema Swelling.

EKG See electrocardiograph and electrocardiogram.

elastin A protein of blood vessels that imparts elasticity.

electrical cardioversion A procedure used by emergency personnel to deliver a strong electrical pulse to a person whose heart has stopped beating. The jolt is supplied by a defibrillator, and is meant to shock the heart back into a normal sinus rhythm.

electrocardiogram (ECG or EKG) The product of an electrocardiograph, it is a printout depicting the heart's electrical activity. An ECG has five parts, each signified with the letter P, Q, R, S, or T, that reflect different phases in the heart activity.

electrocardiograph (ECG or EKG) A device that records the heart's electrical activity as a jagged line on a sheet of paper, which is called an electrocardiogram.

electrolyte A charged particle like calcium (Ca^{2+}) or magnesium (Mg^{2+}) that may have a number of functions in cells.

embryocyst The portion of the blastocyst that eventually develops into the embryo. See trophoblast.

end-diastolic volume The amount of blood in a completely filled ventricle. In an adult, this is typically about 0.12 quarts (120 ml).

endocarditis A medical condition stemming from a bacterial infection of the endocardium.

endocardium The membrane lining the heart.

endometrium The lining of the uterus.

endothelium In blood vessels, it is also known as the tunica intima. The tunica intima forms the innermost layer of blood vessels.

eosinophil A type of granulocyte that appears to be active in the moderation of allergic responses and the destruction of parasites.

epigastric vein A blood vessel of the digestive system. The inferior and superficial epigastric veins flow from the abdominal wall to the external iliac vein or great saphenous vein, respectively.

epinephrine See adrenaline.

epitope The specific area of an antigen to which the B cell receptor binds.

erythroblast An early stage in red blood cell development.

erythroblastosis fetalis See Rh disease.

erythrocytes See red blood cells.

essential thrombocythemia A medical disorder that results in an abundance of blood platelets in the bone marrow.

femoral artery Arising from one of the two external iliac arteries, the femoral artery traverses the thigh to the popliteal artery.

femoral vein A large blood vessel in the thigh that collects blood from the popliteal vein and great saphenous vein and delivers it to the external iliac vein.

fenestrated capillary One of the three types of capillary. It is highly perforated with small openings. See discontinuous capillary and continuous capillary.

fibrinogen A protein in plasma. It functions in blood clotting.

fibular veins See peroneal veins.

foramen ovale An opening between the atria found in the fetal circulation. It typically closes at birth.

fractionation A method for isolating proteins from liquid plasma.

Frank-Starling mechanism Also known as the Frank-Starling relationship. It is the mechanism that maintains blood flow by regulating heart contractility and the related changes in stroke volume.

Frank-Starling relationship See Frank-Starling mechanism.

gastric arteries Blood vessels of the digestive system. The left gastric artery stems from the celiac artery and supplies the stomach and lower part of the esophagus. The right gastric artery stems from the common hepatic artery and eventually connects with the left gastric artery.

gastric veins Blood vessels of the digestive system. Blood from the stomach exits into the gastric veins, which then empty into a number of other veins that ultimately enter the portal vein (in the case of the left and right gastric veins) or the splenic vein (in the case of the short gastric vein).

gastroduodenal artery Blood vessel that separates from the common hepatic artery, and feeds the gastroepiploic and other arteries.

gastroepiploic arteries Blood vessels of the digestive system. The right gastroepiploic artery branches from the gastroduodenal artery. The left gastroepiploic artery branches from the splenic artery. Both provide blood to the stomach and duodenum.

gastroepiploic veins Blood vessels of the digestive system. The left gastroepiploic vein and right gastroepiploic vein drain the stomach into the splenic vein or superior mesenteric vein, respectively.

gemfibrozil A type of drug sometimes used to treat high cholesterol levels.

gestational diabetes A condition of high blood sugar that occurs in 2–4 percent of pregnancies. The temporary condition is typically managed through diet and exercise.

globulins Plasma proteins that function as transportation vehicles for a variety of molecules, in blood clotting and/or in the body's immune responses. They are divided into three types: alpha, beta, and gamma globulins.

glomeruli Clusters of capillaries in the kidneys.

glomerulus The singular form of glomeruli.

glucose A sugar that results mainly from starch digestion.

glycogen A storage form of carbohydrate. The liver converts fats, amino acids, and sugars to glycogen, which functions as a reserve energy supply for the body.

granulocyte The most abundant type of white blood cell. See neutrophil, eosinophil, and basophil.

gray matter A collective term for neurons (nerve cells) in the brain.

great saphenous vein The longest vein in the human body. It delivers blood from the foot to the femoral vein.

hardening of the arteries See atherosclerosis.

HDL See high-density lipoprotein.

heart The muscular pump that powers the circulatory system.

heart attack Also known as a myocardial infarction, this condition happens when the supply of oxygen to a portion of the heart muscle is curtailed to such a degree that the tissue dies or sustains permanent damage.

heart failure A condition in which the heart can no longer carry out its pumping function adequately, resulting in slow blood circulation, poorly oxygenated cells, and veins that hold more blood.

heart murmur A sound produced by the vibration of blood as it traverses a faulty heart valve.

heart transplant The removal of a patient's heart and replacement with a heart from a donor. The medical procedure is used as a last resort to treat heart disease.

heart valves Gateways in the heart that permit blood flow in only one direction.

heme group A ringlike chemical structure that is part of hemoglobin.

hemoglobin A large chemical compound in red blood cells that imparts their red color and also participates in transporting oxygen and carbon dioxide.

hemoglobin C disease A medical condition in which patients have abnormal hemoglobin that crystallizes in red blood cells, deforming them and making them targets for destruction by the spleen.

hemolysis The rupture and destruction of red blood cells.

hemolytic anemia A form of anemia in which the patient's red blood cells are destroyed much faster than they are re-

placed, resulting in an overall loss of erythrocytes.

hemolytic disease of the newborn See Rh disease.

hemophilia A bleeding disorder in which the patient is missing or unable to use certain proteins required for blood clotting.

hemorrhage The loss of blood; usually refers to a significant amount of blood loss in a short time.

hemorrhagic stroke A stroke due to an aneurysm. It may result in bleeding inside the brain or in the space between the brain and the skull.

hepatic arteries The blood vessels supplying the liver and other organs. The common hepatic artery arises from the celiac trunk and supplies the right gastric, gastroduodenal, and proper hepatic arteries. The proper hepatic artery supplies the liver by way of the cystic artery.

hepatic vein The blood vessel that collects blood from the liver and delivers it to the inferior vena cava.

hepatitis An inflammation of the liver, usually due to viral infection. It can be chronic or acute.

hereditary hemorrhagic telangiectasia (HHT) Also called Osler-Weber-Rendu syndrome, it is somewhat similar in its symptoms to hemophilia. Instead of the blood-clotting problem indicative of hemophiliacs, individuals with HHT lack capillaries, so blood moves directly from the arterial system to the venous system.

HHT See hereditary hemorrhagic telangiectasia.

high blood pressure A medical condition that arises when the pressure of the blood against the blood vessel walls exceeds normal limits. It results from a narrowing of the arterioles.

high-density lipoprotein (HDL) Often called the "good" cholesterol. This type of cholesterol curtails the accumulation of low-density lipoprotein in blood vessels.

HIV See human immunodeficiency virus.

hormone A chemical compound, often called a chemical messenger, that the brain and other organs use to communicate with the cells.

human immunodeficiency virus (HIV) The virus responsible for acquired immune deficiency syndrome (AIDS).

hydroxyurea A medication used to treat sickle cell disease.

hyperglycemia A medical condition in which blood-sugar levels are elevated.

hyperkalemia A medical condition in which the patient has a high concentration of the mineral potassium in the circulating blood.

hypertension See high blood pressure.

hypertrophic cardiomyopathy A disease in which the heart walls are thickened, which decreases the volume of the chambers, particularly the left ventricle, and affects the heart's ability to pump effectively.

hypoglycemia A medical condition in which blood-sugar levels are depressed.

hypokalemia A medical condition in which the patient has a low concentration of the mineral potassium in the circulating blood.

hypovolemic shock Shock brought on by blood or fluid loss.

ileocolic artery One of several blood vessels arising from the mesenteric arteries and supplying the intestinal system.

iliac arteries These arise at the end of the abdominal artery. The abdominal artery bifurcates into two common iliac arteries, each of which soon divides again into internal and external iliac arteries.

iliac veins Blood from the femoral vein collects in the external iliac vein, which joins the internal iliac vein and carries blood from the pelvis to form the common iliac vein.

immune thrombocytopenic purpura (ITP) A medical disorder in which antibodies attack and destroy the platelets that are necessary for blood clotting.

immunoglobulins (Ig) Plasma proteins that act as antibodies. The five main types are IgA, IgD, IgE, IgG, and IgM.

incompetent valve A heart valve that closes incompletely.

innominate artery See brachiocephalic artery.

innominate vein See brachiocephalic veins.

inotropic agents Medications used to boost the force of the heart's contractions, thus increasing blood pressure and helping to ensure sufficient oxygenation to body tissues. This may be used to treat heart failure and other conditions.

in-series blood circulation Also known as portal circulation. It is blood flow that travels from one organ to another in series.

insulin A hormone secreted by the pancreas. It allows the body cells to use energy, specifically glucose.

interferons A family of drugs used to regulate the body's immune system. They may be used for such diseases as multiple sclerosis or cirrhosis of the liver.

interlobar arteries Blood vessels that branch from the renal artery to disperse blood throughout the kidney and to glomeruli.

interstitial space The fluid-filled extracellular area.

intestinal artery The blood vessel arising from the superior mesenteric artery and feeding portions of the intestine.

intestinal villi Tiny projections that line the inside wall of the small intestine and the uptake of nutrients by capillaries.

involuntary muscle See smooth muscle.

iron deficiency anemia A form of anemia caused by an insufficient supply of iron. Iron is required to make hemoglobin molecules that carry oxygen in the blood, so an iron deficiency affects oxygen delivery.

ischemic stroke Also known as cerebral infarction. This is a medical condition that occurs when the blood flow of an artery in the brain is disrupted and the affected part of the brain begins to die. Most of the time, the disruption is due to a blood clot that forms in the artery itself or breaks loose from another part of the body and travels to the brain, where it becomes lodged.

isoproterenol A drug that may be prescribed to activate the heart muscle. It may be used to treat bradycardia or asystole.

ITP See immune thrombocytopenic purpura.

jaundice Yellowing of the skin.

jugular veins Blood vessels of the head and/or neck. The anterior jugular vein collects blood from veins of the lower face, traverses the front of the neck, and delivers the blood to the external jugular vein. The external jugular vein is a large vein that also receives blood from within the face and around the outside of the cranium, and empties into one of several veins, including the internal jugular. The internal jugular vein is the largest vein of the head and neck, and also drains blood from the brain and neck. It joins the subclavian vein to form the brachiocephalic vein.

juvenile diabetes Also known as type 1 diabetes. See diabetes. Type 1 diabetes is an autoimmune disease that affects the body's ability to make insulin. It usually appears before the individual reaches 25 years old.

laser angioplasty A medical procedure in which a catheter is threaded through the blood vessels and to a narrowed vessel where a laser is used to clear plaque.

LDL See low-density lipoprotein.

left-ventricular assist device Temporary mechanical implant sometimes used to help pump blood in a person who experiences heart failure. The device is often employed to maintain the patient's heart function until a compatible heart-transplant donor heart is located.

leukemia A blanket term applied to various cancers of the white blood cells.

leukocytes See white blood cells.

low-density lipoprotein (LDL) Often called the "bad" cholesterol. This type of cholesterol can build up on blood-vessel walls and cause health problems.

lumbar veins Blood vessels of the digestive system. Lumbar veins collect blood from the abdominal walls, and deliver it to other veins, including the inferior vena cava.

lumen The internal diameter of a blood vessel. It represents the open space in the vessel through which the blood flows.

Lyme disease A parasitic infection that sometimes leads to painful arthritis and facial paralysis.

lymph Fluid in the vessels of the lymphatic system. It is the interstitial fluid that exits the capillaries and enters surrounding cells during the capillaries' exchange function.

lymphatic system A series of vessels that shunts excess tissue fluid into the veins.

lymph node Filters that separate from lymph any invading organisms and other foreign materials.

lymphocyte A type of leucocyte that detects antigens and serves in the body's immune response. The two main types are B cells and T cells.

macrophage White blood cells that ingest and digest bacteria, other foreign organisms, platelets, and old or deformed red blood cells.

magnetic resonance imaging (MRI) Diagnostic tool for viewing blood flow and locating sites of blood-flow blockage.

malaria A usually acute disease that causes chills and severe fever. A person with the mutation for sickle cell anemia has protection against malaria.

menopause The time at which a woman's monthly menstrual cycle permanently ends. It commonly occurs from about 50 to 55 years of age. See menstruation.

menstruation Sloughing of the endometrium. It typically occurs once a month in females who are of childbearing age. The cycle ceases during pregnancy and ends permanently at menopause.

mesenteric arteries Blood vessels of the digestive system. The inferior and superior mesenteric arteries arise from the abdominal aorta and flow into numerous arteries of the large and small intestines, and the rectum.

mesenteric veins Blood vessels of the digestive system. The superior mesenteric vein drains the small intestine, and the inferior mesenteric collects blood from the colon and rectum. Both deliver their blood to the splenic vein.

metarterioles Small arterioles that lie adjacent to capillaries and may participate in the exchange between the blood and tissues that is normally limited to capillaries.

microvilli Small outgrowths covering the intestinal villi. They increase the surface area of the villi, aiding in nutrient uptake by capillaries.

mini-strokes See transient ischemic attacks.

mitosis A process by which cells duplicate themselves. The parent cell divides to produce two new "daughter" cells.

mitral stenosis A smaller-than-typical mitral valve, often 10 to 50 percent of normal size.

mitral valve Also known as the bicuspid mitral valve. It is a two-cusped heart valve located between the two left chambers of the heart.

molecular weight The combined atomic masses of all atoms in a molecule.

monocyte A type of white blood cell. They become macrophages, large cells that engage in phagocytosis.

mononucleosis Also known as "mono." It occurs when the circulating blood has an abundance of abnormal, mononuclear white blood cells.

MS See multiple sclerosis.

multiple sclerosis (MS) A chronic disease in which the immune system mistakes the insulation around nerve fibers in the central nervous system, including the brain and spinal cord, for foreign material and destroys it.

myelin sheath The insulation around nerve fibers in the central nervous system, including the brain and spinal cord.

myelofibrosis A medical disorder that allows bone marrow cells to reproduce in excess.

myocardial infarction See heart attack.

myocarditis An inflammation of the myocardium. In extreme cases, it can lead to cardiac failure.

myoglobin A chemical released by dead cardiac muscle cells. Its presence is indicative of a heart attack.

nephron The filtering unit of the kidney.

neutrophil The most common type of granulocyte. They are a main bodily defense mechanism against infection, and are particularly suited to engulfing and destroying bacteria, although they can also combat other small invading organisms and materials.

nicotinic acid A substance sometimes used as a drug to treat high cholesterol levels.

nitrates Medications, including nitroglycerin, that dilate blood vessels to improve blood flow. Nitrates are typically used in heart-attack treatment.

nitroglycerin One of several nitrate drugs.

nonsteroidal anti-inflammatory drugs (NSAIDs) Medications used to relieve pain.

NSAIDs See nonsteroidal anti-inflammatory drugs.

nuclear ventriculography Also known as radionuclide ventriculography. It is a diagnostic procedure used especially in heart patients to view the heart and major blood vessels.

Osler-Weber-Rendu syndrome See hereditary hemorrhagic telangiectasia.

osmosis A process that seeks to equalize the water-to-solute ratio on each side of a water-permeable membrane.

ovarian vein One of a pair of veins serving the female reproductive system.

pacemaker See SA node.

parasympathetic nervous system Also known as the vagal system. It is one of two major divisions of the autonomic nervous system. It functions to inhibit the pacemaker and lower the heart rate. See sympathetic nervous system.

paroxysmal cold hemoglobinuria (PCH) Also called Donath-Landsteiner syndrome. This is a blood disorder in which exposure to cold temperatures prompts antibodies and a cell-destroying protein to seek out and exterminate red blood cells, particularly those moving through the hands and feet.

partial pressure Within the circulatory system, it is a term used to describe the relative oxygen concentration in tissues. For example, hemoglobin has a differential ability to bind oxygen: It picks up oxygen when the partial pressure in surrounding tissues is high, as it is in the lungs, and drops off oxygen when the partial pressure in the surrounding tissues is low, as it is in the tissues.

PCH See paroxysmal cold hemoglobinuria.

penis The organ in the male reproductive system that delivers sperm.

pericarditis A medical condition stemming from an infection of the pericardium.

pericardium The two-layered membranous sac around the heart.

pericytic venules See postcapillary venules.

pernicious anemia A form of anemia in which patients are unable to absorb vitamin B_{12}, which is necessary for the production of red blood cells.

peroneal veins Also known as fibular veins. They drain the lower leg and ankle, and deliver the blood to the posterior tibial vein.

petechiae Small, reddish-purple spots on the skin, in the nose, inside the mouth, and in other mucus membranes. It may occur as a symptom of immune thrombocytopenic purpura.

phagocytosis The process of engulfing and destroying bacteria and other antigens.

pH level A measure of acidity or alkalinity based on the concentration of hydrogen ions (H^+).

pituitary gland A gland, located at the base of the brain behind the eyes, that produces a variety of hormones.

placenta In a pregnant female, this structure links the mother to the umbilical cord, which then attaches to the fetus.

plaque Deposits of fat, cholesterol, and other materials build up along the inside of blood vessels, and can narrow the lumen of the vessels.

plasma The liquid portion of blood in which red and white blood cells, platelets, and other blood contents float.

plasminogen A beta globulin that participates in blood clotting.

platelets Also known as thrombocytes. They are round or oblong disks in the blood that participate in blood clotting.

pluripotent hemopoietic (blood-forming) stem cells Undifferentiated cells that have the genetic potential to mature into red blood cells, white blood cells, or platelets.

polycythemia vera A medical disorder that allows bone marrow cells to reproduce in excess.

popliteal artery A blood vessel that arises from the femoral artery and traverses the knee before dividing into the posterior and anterior tibial arteries.

popliteal vein A blood vessel that collects blood from the anterior and posterior tibial veins, and empties into the femoral vein.

portal circulation See in-series blood circulation.

portal vein A blood vessel that arises from the splenic vein and superior mesenteric vein. It empties into the liver.

postcapillary venules Also known as pericytic venules. Small venules that lie adjacent to capillaries and may participate in the exchange between the blood and tissues that is normally limited to capillaries.

probucol A type of drug sometimes used to treat high cholesterol levels.

protein Complex chemical compounds that are essential to life.

prothrombin A beta globulin that participates in blood clotting.

protozoa Single-celled, eukaryotic organisms, including many parasites.

pudendal veins Blood vessels of the lower abdominal and pelvic region. The external pudendal vein drains the lower abdomen and external genitals, and flows into the great saphenous or femoral veins. The internal pudendal vein drains the genital region and flows into the internal iliac vein.

pulmonary artery The blood vessel that originates at the right ventricle, then splits into two branches. The left and right pulmonary arteries lead to the left and right lung, respectively.

pulmonary circulation The transit of blood from the heart to the lungs and back to the heart. Blood picks up oxygen and drops off carbon dioxide in this circulatory route.

pulmonary semilunar valve The three-cusped heart valve located between the right ventricle and pulmonary artery.

pulmonary veins Four blood vessels that flow from the lungs to the left atrium.

Purkinje fibers A mesh of modified muscle fibers located in the base of the ventricle wall. The fibers receive the electrical impulse from the bundle of His and deliver it to the ventricle, which then contracts.

radial artery A blood vessel in each lower arm that receives blood from the brachial artery and delivers it to numerous arteries of the forearm, wrist, and hand.

radial vein A blood vessel in each arm that collects blood from veins in the hand. It eventually merges with the ulnar vein into the brachial vein.

radionuclide ventriculography See nuclear ventriculography.

Raynaud's disease Also known as Raynaud's phenomenon. It is a medical condition in which cold temperatures trigger the arteries to contract, particularly in the extremities, and limit blood flow to these areas.

Raynaud's phenomenon See Raynaud's disease.

rectal vein Blood vessels in the digestive system that drain parts of the rectum. The inferior rectal vein joins the internal pudendal vein, which flows into the internal iliac vein, while the middle rectal vein connects directly to the internal iliac vein. The superior rectal vein flows directly into the inferior mesenteric vein.

red blood cells Also known as erythrocytes. These are the cells in the blood that are responsible for gathering and delivering oxygen and nutrients to the body tissues, and for disposing of the tissue's waste products.

renal artery A pair of blood vessels that arise from the abdominal aorta. Each feeds a kidney and adrenal gland, and the ureter.

renal veins A pair of blood vessels that drain the two kidneys. They empty into the inferior vena cava.

resistance vessels Another name for arterioles. The term reflects their function in regulating the amount of blood entering the capillaries.

respiratory pigment A term sometimes used to describe hemoglobin.

restrictive cardiomyopathy A type of cardiomyopathy, in which abnormal tissue causes the ventricles to stiffen and affects their ability to adequately pump blood.

reticulocyte The developmental stage preceding a fully formed red blood cell.

Rh disease Also known as hemolytic disease of the newborn or erythroblastosis fetalis. It involves the destruction of fetal red blood cells brought on by the mother's immune response.

rheumatic fever An inflammatory disease that can result in damage to heart valves.

Rh factor Also known as D antigen. A protein present on the surface of red blood cells in some but not all people. Those with the protein are said to have Rh positive blood, and those without the protein have Rh-negative blood.

SA node Also known as the sinoatrial node, or pacemaker. This is a group of small and weakly contractile modified muscle cells that spontaneously deliver the electrical pulses that trigger the heart's contraction.

semilunar valves Valves, which are shaped like half-moons, that ensure blood movement in only one direction. They are found in the heart and in large blood vessels.

septum The thick muscular wall that separates the right and left sides of the heart.

shock A medical condition in which the patient experiences a sudden drop in blood flow.

sickle cell anemia A medical condition resulting from a slightly different hemoglobin, called Hb S, that causes red blood cells to become crescent-shaped when they give up their oxygen. These sickled cells can clump together to block blood vessels and to accumulate in and damage organs. In addition, sickle cells die much sooner than normal red blood cells, resulting in anemia.

sigmoidal artery A blood vessel that arises from the inferior mesenteric artery and supplies blood to the lower abdominal region.

sinoatrial node See SA node.

sinusoidal capillary See discontinuous capillary.

smooth muscle Also known as an involuntary muscle. It is a type of muscle that is controlled by the autonomic nervous system, rather than by willful command, as is the striated muscle.

solutes Dissolved materials.

spermatic vein One of a pair of veins serving the male reproductive system.

spherocytosis Also known as congenital sperocytic anemia. A medical condition that affects the membranes of red blood cells and changes them from their normal disk shape to spheres. The spleen detects the abnormal cells and destroys them.

sphygmomanometer Also known as a blood-pressure cuff. It is a tool used by the medical community to determine a patient's blood pressure.

splenic artery Blood vessel that arises from the celiac trunk and branches into numerous arteries that feed the stomach and peritoneum, pancreas, and spleen.

splenic vein A large blood vessel that collects blood from the spleen. It joins the superior mesenteric vein to create the portal vein.

splenomegaly An enlargement of the spleen. Sometimes, this medical condition may become so extreme that the spleen ruptures and begins to release blood into the abdomen.

statins A group of drugs sometimes used to treat high cholesterol levels.

stem cells Undifferentiated cells. They have the genetic potential to mature into specific cell types. Some stems are only able to become one type of cell, while others have the ability to become any number of different cells. See pluripotent hemopoietic (blood-forming) stem cells.

stenosis of a valve A severe narrowing of a heart valve.

stent Tiny, tube-shaped, metal devices that are inserted by a catheter as part of a medical procedure to open narrowed blood vessels. They remain in place at the problem site to hold the vessel open.

sternum Breastbone.

striated muscle Also known as voluntary muscle. A person can consciously control the action of striated muscle.

stroke Often called a "brain attack" or cerebrovascular disease, a stroke is a sudden blockage of blood flow to a part of the brain. It can result in brain cell-death, and loss of functions that are controlled by those cells.

stroke volume Also known as total cardiac output. It is the amount of blood pumped from the heart into the aorta in one contraction.

subclavian arteries Blood vessels that supply the arms, much of the upper body, and the spinal cord. The right subclavian artery branches from the brachiocephalic artery, while the left divides off of the aortic arch. Numerous arteries arise from each.

subclavian veins Primary blood vessels draining the arms. They collect blood from the axillary vein and later merge with the internal jugular vein to produce the brachiocephalic vein.

sympathetic nervous system One of two major divisions of the autonomic nervous system. It functions to stimulate the pacemaker and boost the heart rate. See parasympathetic nervous system.

systemic circulation The transit of blood from the heart to the body (except the lungs) and back to the heart. See pulmonary circulation and coronary circulation.

systole The heart's contractile period.

tachycardia A fast-beating arrhythmia.

T cells Also known as T lymphocytes. They are one of two main types of lymphocyte, and participate in the body's immune response.

terminal arterioles Arterioles that feed capillaries.

thalassemia A form of anemia in which patients are unable to properly synthesize hemoglobin.

thrombocytes See platelets.

thrombolytic agents Often described as "clot-busting" drugs, they include streptokinase, urokinase, antistreplase, and tissue plasminogen activator (see stroke). Thrombolytic agents, which employ different methods of breaking up blood clots, are typically prescribed to treat heart attacks.

thromboplastin A substance released by damaged tissue and platelets. With calcium, it promotes the formation of blood clots.

thrombus A blood clot that forms in the heart or in a blood vessel and remains stationary. If it moves, it is termed an embolus.

tibial arteries Blood vessels of the lower leg. The posterior and anterior tibial arteries arise from the popliteal artery and supply blood to arteries feeding the lower leg, ankle, and foot.

tibial veins Blood vessels of the lower leg. The anterior and posterior tibial veins drain the leg, then join together to form the popliteal vein.

tissue plasminogen activators (tPA) A group of drugs used particularly in the treatment of stroke and heart attack. Known as "clot-busters," they dissolve clots and restore blood flow.

T lymphocytes See T cells.

total cardiac output See stroke volume.

tPA See tissue plasminogen activators.

transcranial doppler Diagnostic tool for viewing blood flow and locating sites of blood-flow blockage.

transient ischemic attacks (TIAs) Also known as mini-strokes, these are temporary strokes in which the precipitating blood clot dissolves before it has caused a long-enough stoppage of blood flow to

bring about lasting effects. TIAs may or may not precede major strokes.

tricuspid atrioventricular valve See AV valve.

trophoblast The portion of the blastocyst that forms the passageway between the fetus and mother. See embryocyst.

troponin A chemical released by dead cardiac muscle cells. Its presence is indicative of a heart attack.

tunica adventitia Fibrous connective tissue forming the outer of the three layers comprising arteries, arterioles, veins, and venules. See tunica intima and tunica media.

tunica intima Also known as endothelium. It forms the innermost of the three layers comprising arteries, arterioles, veins, and venules. Capillaries are composed of only a single layer of endothelial cells. See tunica adventitia and tunica media.

tunica media Muscular and elastic tissue forming the middle of the three layers comprising arteries, arterioles, veins, and venules. See tunica adventitia and tunica intima.

type AB blood Blood containing antigens called "A" and "B." A person with type AB blood can receive blood donations of type AB, type O, type A, or type B.

type A blood Blood containing a certain antigen called "A." Due to potential antigen reactions, a person with type A blood can receive blood donations of type A and type O, but not type B or type AB.

type B blood Blood containing a certain antigen called "B." Due to potential antigen reactions, a person with type B blood can receive blood donations of type B and type O, but not type A or type AB.

type O blood Blood containing neither of the antigens called "A" and "B." Due to potential antigen reactions, a person with type O blood can receive blood donations of type O, but not type A, type B, or type AB.

ulnar artery A blood vessel in each lower arm that receives blood from the brachial artery and delivers it to numerous arteries of the forearm, wrist, and hand.

ulnar vein A blood vessel in each arm that collects blood from veins in the hand. It eventually merges with the radial vein into the brachial vein.

umbilical arteries Two blood vessels in a fetus connecting to capillaries in the fetal villi, which in turn connect with maternal capillaries.

umbilical cord With the placenta, it forms the passageway between mother and fetus.

umbilical vein The blood vessel in the fetal circulation that collects blood from capillaries in the placenta and delivers it to either the liver or the ductus venosus.

urea A solid waste product that forms as cells metabolize proteins.

ureter A long tube that delivers urine from the kidney to the bladder.

uterine veins A pair of blood vessels that collect blood from the uterus. They merge with the internal iliac vein.

uterus The organ in a female that holds and nourishes the developing fetus.

vagal system See parasympathetic nervous system.

vagina Female reproductive opening.

valvular regurgitation Leakage against the normal flow of blood through a heart valve.

valvular stenosis Narrowing of the heart valve.

vasculature See blood vessels.

vasoconstrictor nerves Nerves that signal the veins to constrict.

vasodilator One of a group of medications that dilate blood vessels.

vaso vasorum Small blood vessels inside the walls of large blood vessels. They provide nourishment to the thick vessel walls.

vein A blood vessel that transports oxygen-depleted blood from body tissues back to the heart. Veins are larger than the venules, which also transport blood back to the heart.

vena cava One of two large veins, the superior and inferior venae cavae, bringing blood from the body back to the heart.

ventricle In the human heart, it is one of the heart's two lower chambers.

venule A smaller blood vessel that transports oxygen-depleted blood from body tissues back to the heart. Venules are smaller than veins, which also transport blood back to the heart.

vertebral arteries A pair of blood vessels on each side of the neck that arise from the subclavian arteries. They unite at the basilar artery.

villi See intestinal villi.

viral myocarditis A common form of the medical condition myocarditis.

viscosity The thickness of a liquid.

viscous Thick, or slow flowing.

voluntary muscle See striated muscle.

von Willebrand disease A medical condition in which patients are deficient in blood-clotting factors. This leads to uncontrolled bleeding.

West Nile fever A viral infection. Humans are infected following a bite from an infected mosquito. The mosquitoes, in turn, become infected after they bite a bird that carries the virus.

white blood cells Also known as leukocytes. These are the cells in the blood that function in the body's defense mechanism to detect, attack, and eliminate foreign organisms and materials.

Organizations and Web Sites

Adult Congenital Heart Association
273 Perham Street
West Roxbury, MA 02132
Phone: (617) 325-1191
www.achaheart.org

Information and support for adolescents through adults who have congenital heart disease, and their families. Forums and links to local support groups. Also an information source for healthcare professionals.

American Association of Blood Banks
8101 Glenbrook Road
Bethesda, MD 20814-2749
Phone: (301) 907-6977
Fax: (301) 907-6895
Email: aabb@aabb.org
www.aabb.org

Designed for those who are interested in blood banks and transfusion-medicine activities. Includes information about donor screening, transfusion-transmitted diseases, and a historical timeline.

American Association for Thoracic Surgery
900 Cummings Center, Suite 221-U
Beverly, MA 01915
Phone: (978) 927-8330
Fax: (978) 524-8890
Email: aats@prri.com
www.aats.org

For medical professionals, the site provides links to journals and to educational and training information.

American Cancer Society
P.O. Box 102454
Atlanta, GA 30368-2454
Phone: (800) ACS-2345
www.cancer.org

Information on all types of cancer. Patients, survivors, and their families, as well as healthcare professionals, will find news, updates on clinical trials, links to support groups, and other resources.

American College of Cardiology
Heart House
9111 Old Georgetown Road
Bethesda, MD 20814-1699
Phone: (800) 253-4636, ext. 694 or (301) 897-5400
Fax: (301) 897-9745
www.acc.org

An advocacy and educational site for medical professionals who are treating cardiovascular patients.

American Heart Association
National Center
7272 Greenville Avenue
Dallas, TX 75231
Phone: (800) 242-8721
www.americanheart.org

A comprehensive site with information on prevention, warning signs, and treatment of heart disease and stroke. Also provides links to numerous publications, to local branches of the association throughout the country, and to the American Stroke Association, which is a division of the AHA.

American Sickle Cell Anemia Association
10300 Carnegie Avenue
Cleveland Clinic/East Office Building (EEb18)
Cleveland, OH 44106
Phone: (216) 229-8600
www.ascaa.org

The Web site features educational materials, support groups, events, and a message board.

American Society of Echocardiography
1500 Sunday Drive, Suite 102
Raleigh, NC 27607
Phone: (919) 861-5574
Fax: (919) 787-4916
www.asecho.org

Designed for medical professionals. Includes recommendations and guidelines for cardiovascular ultrasound.

American Society of Hypertension
148 Madison Avenue, Fifth Floor
New York, NY 10016

Phone: (212) 696-9099
Fax: (212) 696-0711
Email: ash@ash-us.org
www.ash-us.org

Designed for medical professionals. Includes links to publications and continuing education.

American Stroke Association
National Center
7272 Greenville Avenue
Dallas, TX 75231
Phone: (888) 478-7653
www.strokeassociation.org

Includes the warning signs of stroke, a description of stroke, and treatment. Links to local stroke association groups are also provided.

Angioma Alliance
107 Quaker Meeting House Road
Williamsburg, VA 23188
Phone: (757) 258-3355 or (866) 432-5226
Email: Info@Angiomaalliance.org
www.angiomaalliance.org

Topics relating to cavernous angioma, including ongoing research on the condition and the latest research findings. An online patient brochure is also available.

Canadian Adult Congenital Heart Network
Timothy P. Caley, Chair, Adult Congenital Heart Council
Phone: (416) 417-6523
Fax: (416) 340-5014
Email: caley@cachnet.org
www.cachnet.org

For adult patients and their healthcare professionals. Newsletter, message board, educational information, and links to medical sites and other resources.

Cardiology Channel
www.cardiologychannel.com

Information for patients on conditions, diagnostic tests, treatments, and clinical tests, as well as forums and numerous resources.

Cardiology Online
Conceptis Technologies, Inc.
390 Guy Street, Suite 109
Montreal, Québec H3J 1S6 Canada
Phone: (514) 931-5434 or (888) 747-5434
Fax: (514) 931-5362
Email: info@theheart.org
www.theheart.org

For cardiologists, allied healthcare professionals, and referring physicians. Information on the prevention and care of cardiovascular disorders. Registration is required for full use of the site.

Cardiomyopathy Association
40 The Metro Centre
Tolpits Lane
Watford
Herts
WD18 9SB
UK
Phone: +44 (0) 1923 249 977
Fax: +44 (0) 1923 249 987
www.cardiomyopathy.org

The Web site provides support and advice for patients with specific information on different types of cardiomyopathy; also has information for medical professionals.

Centers for Disease Control
Public Inquiries/MASO
Mailstop F07
1600 Clifton Road
Atlanta, GA 30333
Phone: (800) 311-3435
www.cdc.gov

Includes Web pages and numerous publications devoted to heart disease, blood disorders, AIDS/HIV, and other cardiovascular disorders. Many publications, like those on AIDS/HIV, are specific to certain populations, provide detailed statistics, and offer prevention suggestions.

Children's Hemiplegia and Stroke Association
Suite 305, PMB 149
4101 W. Green Oaks
Arlington, TX 76016
Phone: (817) 492-4325
Email: info437@chasa.org
www.chasa.org

Offers support and information to families of infants through young adults who have various conditions, including those who have experienced prenatal, childhood, infant, perinatal, neonatal, or in utero stroke.

Cleveland Clinic Heart Center
Phone: (216) 445-9288 or (866) 289-6911
www.clevelandclinic.org/heartcenter

Comprehensive patient information about the heart and vasculature, including a history of cardiovascular innovations. Also provides information for medical professionals.

Congenital Heart Center Encyclopedia: A Service of Children's Hospital Medical Center of Cincinnati
Cincinnati Children's Hospital Medical Center
3333 Burnet Avenue
Cincinnati, OH 45229-3039

Phone: (513) 636-4432 or (513) 636-4770
Email: thc@cchmc.org
www.cincinnatichildrens.org/heartcenter/encyclopedia

Encyclopedia of various cardiovascular disorders that affect children. Symptoms, diagnoses, and treatment options are covered.

Cut to the Heart
www.pbs.org/wgbh/nova/heart

Covers anatomy and physiology of the heart, heart disorders and treatment options, and some of the history of heart surgery.

European Society of Cardiology
The European Heart House
2035 Route des Colles
B.P. 179 - Les Templiers
FR-06903 Sophia Antipolis
France
Phone: +33.4.92.94.76.00
Fax : +33.4.92.94.76.01
www.escardio.org

Provides lists of and links to dozens of national cardiology societies from throughout Europe. It also connects to several professional journals, and provides information about educational resources.

"For Your Heart" National Women's Health Information Center
8550 Arlington Boulevard, Suite 300
Fairfax, VA 22031
Phone: (800) 994-9662
Email: 4woman@soza.com
www.4woman.gov/hhs

Designed for women, this site leads users through a short, confidential survey and to heart-related articles selected for the user. The survey need not be taken to reach materials on specific conditions, such as high cholesterol or high blood pressure.

Global Cardiology Network
Email: webmaster@escardio.org
www.globalcardiology.org

Designed for cardiology professionals, the site provides information on clinical trials, continuing medical education, cardiology meetings, and links to member organizations.

Healthfinder
P.O. Box 1133
Washington, DC 20013-1133
Email: healthfinder@nhic.org
www.healthfinder.gov

A U.S. federal Web site that links to information and Web sites from more than 1,700 sources. Visitors type in a word and the site searches the sources for related news articles.

HeartCenterOnline
One South Ocean Boulevard, Suite 201
Boca Raton, FL 33432
Fax: (561) 620-9799
www.heartcenteronline.com

With separate entrances for patients and medical professionals, this site offers information on conditions and diseases affecting the cardiovascular system. The patient section yields guides to arrhythmias, heart attack, angina, and other disorders, and cardiologist-edited articles on specific procedures, including treatment devices like catheters and pacemakers.

Heart Info
www.heartinfo.org

A patient-oriented site with articles on recent research, treatments, and descriptions of numerous disorders. It also includes nutrition and fitness guides, and an ask-the-doctor service that requires users to sign in.

Heart Preview Gallery, The Franklin Institute Online, The Franklin Institute Science Museum
www.fi.edu/biosci/preview/heartpreview.html

Audio and video clips take visitors through various cardiovascular procedures, including an echocardiography and open heart surgery. Through animations, visitors can also learn how to take their own pulse or see how blood circulates. It also includes teacher resource materials that can be integrated into the school curriculum.

Hereditary Hemorrhagic Telangiectasia Foundation
HHT Foundation International, Inc.
P.O. Box 329
Monkton, MD 21111
Phone: (800) 448-6389 or (410) 357-9932
Fax: (410) 357-9931
Email: hhtinfo@hht.org
www.hht.org/web/

For patients, their families, and their doctors, this site includes news about the disease, access to a newsletter, and a list of treatment sites.

Hypertrophic Cardiomyopathy Association
P.O. Box 306
Hibernia, NJ 07842
Phone: (973) 983-7429 8am to 8pm Eastern time ONLY
Email: support@4hcm.us
www.hcma-heart.com

Describes the symptoms, diagnosis, and treatment of this condition, as well as links to related sites.

International Society for Adult Congenital Cardiac Disease
1500 Sunday Drive, Suite 102
Raleigh, NC 27607
Phone: (919) 861-5578

Fax: (919) 787-4916
Email: info@isaccd.org
www.isaccd.org

For healthcare professionals and their adult and adolescent patients. News, information, and Web resources.

Leukemia and Lymphoma Society
Home Office
1311 Mamaroneck Avenue
White Plains, NY 10605
Phone: (914) 949-5213 or (800) 955-4572
Fax: (914) 949-6691
www.leukemia-lymphoma.org

Disease information along with patient services, such as information on treatments and support groups. Offers numerous downloadable documents.

Med Help Heart Forum
www.medhelp.org/forums/cardio/wwwboard.html

This site is a forum filled with questions from patients and their family and friends, as well as answers from doctors at the Cleveland Clinic Heart Center. Topics include all manner of cardiovascular problems.

Mended Hearts, Inc.
7272 Greenville Avenue
Dallas, TX 75231-4596
Phone: (214) 706-1442
Fax: (214) 706-5245
Email: info@mendedhearts.org
www.mendedhearts.org

Provides a variety of education resources for heart-disease patients and their families.

National Center for Health Statistics
3311 Toledo Road
Hyattsville, MD 20782
Phone: (301) 458-4000
www.cdc.gov/nchs

Visitors may use the search tool to find a wide variety of U.S. health statistics, including data regarding cardiovascular disorders and conditions.

National Heart, Lung and Blood Institute
NHLBI Health Information Center
Attention: Web Site
P.O. Box 30105
Bethesda, MD 20824-0105
Phone: (301) 592-8573
Fax: (301) 592-8563
Email: nhlbiinfo@nhlbi.nih.gov
www.nhlbi.nih.gov

Offers information for patients and the public, medical professionals, and researchers. For patients and the public, numerous publications and fact sheets on various conditions are available.

National Hemophilia Foundation
116 West 32nd Street, 11th Floor
New York, NY 10001
Phone: (212) 328-3700
Fax: (212) 328-3777
www.hemophilia.org

Visitors may order numerous publications (individual copies are typically free for members) about hemophilia. These include materials on how parents and school teachers and others care for a child with the disorder, and manuals for healthcare professionals.

National High Blood Pressure Education Program, NIH
NHLBI Health Information Center
Attention: Web Site
P.O. Box 30105
Bethesda, MD 20824-0105
Phone: (301) 592-8573
Fax: (301) 592-8563
Email: nhlbiinfo@nhlbi.nih.gov
www.nhlbi.nih.gov/about/nhbpep/

Includes numerous downloadable patient-oriented publications on controlling and preventing high blood pressure, and educational materials for medical professionals.

National Institute of Neurological Disorders and Stroke
NIH Neurological Institute
P.O. Box 5801
Bethesda, MD 20824
Phone: (800) 352-9424 or (301) 496-5751
www.ninds.nih.gov

Includes information on numerous conditions, including multiple sclerosis and stroke. Information runs the gamut from a description of the disorder and its prognosis to treatment options and current research.

National Stroke Association
9707 E. Easter Lane
Englewood, CO 80112
Phone: (303) 649-9299 or (800) STROKES
Fax: (303) 649-1328
www.stroke.org

Information on stroke for patients and medical professionals, including a professional resource center, articles from *Stroke Smart* magazine, and a list of other stroke resources.

Nobel e-Museum
Email: info@nobel.se
www.nobel.se

Provides information on Nobel Prize winners, including biographies of prize winners and details of their contributions.

North American Society of Pacing and Electrophysiology
6 Strathmore Road
Natick, MA 01760-2499
Phone: (508) 647-0100
Fax: (508) 647-0124
www.naspe.org

Resources for patients and medical professionals dealing with arrhythmias. A Patient Information Center details symptoms, the condition, and treatment options.

PediHeart Organization
Email: webmaster@Pediheart.com
www.pediheart.org

Designed for children with heart disease, their parents, and their care professionals. In the section called KidZone, children can learn about their condition, post a Web page, or ask a question.

Sickle Cell Disease Association of America
Sickle Cell Disease Association of America, Inc.
200 Corporate Pointe, Suite 495
Culver City, CA 90230-8727
Phone: (800) 421-8453
Fax: (310) 215-3722
Email: scdaa@sicklecelldisease.org
www.sicklecelldisease.org

Considerable patient information, including treatment options and new research, and a number of message boards targeting community leaders, patients, and family members who want to ask questions, offer answers, or get involved in advocacy issues.

Society of Thoracic Surgeons
633 N. Street Clair Street, Suite 2320
Chicago, IL 60611-3658
Phone: (312) 202-5800
Fax: (312) 202-5801
Email: sts@sts.org
www.sts.org

Although much of the site is for medical professionals, it also provides an index of patient-oriented articles dealing with the cardiovascular system. Included among these are comprehensive articles on valve replacement and coronary artery bypass grafting.

Stroke Awareness for Everyone
Stroke Awareness for Everyone, Inc.
8906 E. 96th Street, #311
Fishers, IN 46038
Fax: (317) 585-9563

Email: aboutSAFE@strokesafe.org
www.strokesafe.org

Articles and other materials on stroke and recovery, as well as a free, downloadable
caregiver's handbook. Also lists links to other stroke-related organizations and support groups.

StrokeSurvivors International

www.strokesurvivors.org

Offers an e-mail list group to patients, doctors, and others who are interested in
stroke. A list of stroke-related organizations from around the world is provided.

Sudden Arrhythmia Death Syndromes Foundation

508 E. South Temple, Suite 20
Salt Lake City, UT 84102
Phone: (800) 786-7723
Email: sads@sads.org
www.sads.org

Downloadable newsletters provide information on arrhythmia, upcoming events,
and treatment options. Many articles are also posted, as well as links to related organizations.

Transplant Recipients International Organization, Inc. or TRIO

2117 L Street NW, #353
Washington, DC 20037
Phone: (800) TRIO-386 or (202) 293-0980
Email: triointl@aol.com
www.trioweb.org/

Information on transplantation and organ donation transplant candidates, recipients,
donors, and their families. Contains links to publications dealing specifically with
heart transplants, and to chapters throughout the United States and a few in other
countries.

U.S. Food and Drug Administration

5600 Fishers Lane
Rockville, MD 20857
Phone: (888) 463-6332
www.fda.gov

Offers news about activities of the U.S. Food and Drug Administration, including
regulations on drug and medical devices that may be used in or proposed for use in
the treatment of cardiovascular disorders. For example, its section on HIV/AIDS provides information on clinical trials and drug developments, along with articles and
brochures on policy issues and other topics.

U.S. National Library of Medicine

8600 Rockville Pike
Bethesda, MD 20894
www.nlm.nih.gov

Provides links to such databases as MEDLINE/PubMed, MEDLINEplus, and NLM Gateway, which in turn link to articles and abstracts on a full range of medical topics.

World Federation of Hemophilia
1425 René Lévesque Boulevard W, Suite 1010
Montréal, Québec
H3G 1T7 Canada
Phone: (514) 875-7944
Fax: (514) 875-8916
Email: wfh@wfh.org
www.wfh.org

In addition to considerable news about the disorder and hemophilia-related events, the site contains a wide range of detailed and practical information for patients and their families, plus an ask-the-doctor service.

World Heart Federation
5, avenue du Mail
1205 Geneva - Switzerland
Phone: (+41 22) 807 03 20
Fax: (+41 22) 807 03 39
Email: admin@worldheart.org
www.worldheart.org

Offers information on its activities in education and advocacy, as well as current and past issues of its quarterly newsletter *Heartbeat*.

Bibliography

Aaronson, Philip I., and Jeremy P. T. Ward, with Charles M. Wiener, Steven P. Schulman, and Jaswinder S. Gill. *The Cardiovascular System at a Glance*. Oxford: Blackwell Science Limited, 1999.

"About the Blood-Brain Barrier." http://users.ahsc.arizona.edu/davis/blood.htm.

"Aging Marijuana Smokers Face Risk of Heart Attack." http://www.newswise.com/articles/view/17479/.

"Akutsu III Total Artificial Heart." http://www.texasheartinstitute.org/akutsu.html.

"Alessandro Volta (1745–1827)." http://www.corrosion-doctors.org/Biographies/VoltaBio.htm.

"Alfred Blalock." http://www.whonamedit.com/doctor.cfm/2036.html.

Avraham, Regina. *The Circulatory System*. Philadelphia: Chelsea House Publishers, 2000.

"B Cells and T Cells." http://users.rcn.com/jkimball.ma.ultranet/BiologyPages/B/Blood.html.

Bellis, M. "Early Heart Pacemaker." http://inventors.about.com/library/inventors/blcardiac.htm.

Berne, Robert M., and Matthew N. Levy. *Cardiovascular Physiology*, 6th ed. St. Louis: C. V. Mosby–Year Book, 1992.

"Blood Brain Barrier." apu.sfn.org/content/Publications/BrainBriefings/blood-brain.html.

Bowden, Mary Ellen. "Paul Ehrlich (1854–1915)." http://www.chemheritage.org/EducationalServices/pharm/chemo/readings/ehrlich/pabio.htm.

Burke, Allen P., Russell P. Tracy, Frank Kolodgie, Gray T. Malcom, Arthur Zieske, Robert Kutys, Joseph Pestaner, John Smialek, and Renu Virmani. "Elevated C-Reactive Protein Values and Atherosclerosis in Sudden Coronary Death: Association With Different Pathologies." *Circulation* 105 (April 2002): 2019–2023.

Chobanian, Aram V., George L. Bakris, Henry R. Black, William C. Cushman, Lee A. Green, Joseph L. Izzo Jr., Daniel W. Jones, Barry J. Materson, Suzanne Oparil, Jackson T. Wright Jr., and Edward J. Roccella. "The Seventh Report of the Joint

National Committee on Prevention, Detection, Evaluation, and Treatment of High Blood Pressure: The JNC 7 report." *Journal of the American Medical Association* 289 (21 May 2003): 2560–2571.

"Cholesterol." http://nhlbisupport.com/chd1/how.htm.

"Cholesterol Levels." http://www.americanheart.org/presenter.jhtml?identifier=987.

Christensen, Benedicte, Annhild Mosdol, Lars Retterstol, Sverre Landaas, and Dag S. Thelle. "Abstention from Filtered Coffee Reduces the Concentrations of Plasma Homocysteine and Serum Cholesterol—A Randomized Controlled Trial." *American Journal of Clinical Nutrition* 74, no. 3 (September 2001): 302–307.

Christensen, Damarias. "Teasing Out Tea's Heart-Healthy Effect." *Science News* 158 (2 December 2000): 366.

———. "Things Just Mesh: Making Stents Even Better at Keeping Arteries Open." *Science News* 160 (24 November 2001): 328–330.

———. "Walking and Eating for Better Health." *Science News* 160 (8 September 2001): 159.

Clarke, Sarah C., Peter M. Schofield, Andrew A. Grace, James C. Metcalfe, and Heide L. Kirschenlohr. "Tamoxifen Effects on Endothelial Function and Cardiovascular Risk Factors in Men with Advanced Atherosclerosis." *Circulation* 103 (March 2001): 1497–1502.

Clendening, Logan. *Source Book of Medical History.* New York: Dover Publications, 1942.

"Cocaine Can Triple Risk of Aneurysm." http://www.newswise.com/articles/view/15992/.

"Concord Grape Juice May Reduce Blood Pressure in Hypertensive Men." http://www.newswise.com/articles/view/35058/.

Cook, P. J., G.Y.H. Lip, P. Davies, D.G. Beevers, R. Wise, and D. Honeybourne. "*Chlamydia pneumoniae* Antibodies in Severe Essential Hypertension." *Hypertension* 31 (February 1998): 589–594.

Dake, M.D., N. Kato, R.S. Mitchell, C.P. Semba, M.K. Razavi, T. Shimono, T. Hirano, K. Takeda, I. Yada, and D.C. Miller. "Endovascular Stent–Graft Placement for the Treatment of Acute Aortic Dissection." *New England Journal of Medicine* 340 (20 May 1999): 1546–1552.

De Jauregui, Ruth. *100 Medical Milestones that Shaped World History.* San Mateo, CA: Bluewood Books, 1998.

Des Jardins, Terry. *Cardiopulmonary Anatomy and Physiology: Essentials for Respiratory Care,* 3rd ed. Albany, NY: Delmar Publishers, 1998.

"Diagram of Venous Drainage of the Intestine." http://www.umassmed.edu/colonic_neoplasia/coca_02a.html.

"Diet and Salt Intake, Reducing Blood Pressure." http://www.newswise.com/articles/view/18895/.

"Diet, Exercise Together Effective in Controlling High Blood Pressure." Duke University. http://dukemednews.duke.edu/news/article.php?id=6507.

"Dr. Maude Elizabeth Seymour Abbott." http://www.cdnmedhall.org/Inductees/abbott_94.htm.

"Dr. Wilfred Bigelow." http://www.utoronto.ca/bantresf/HallofFame/Bigelow.html.

"*ECG* Library: A Brief History of Electrocardiography." http://www.sh.lsuhsc.edu/fammed/OutpatientManual/EKG/ecghist.html.

"Ehrlich Finds Cure for Syphilis." http://www.pbs.org/wgbh/aso/databank/entries/dm09sy.html.

"Ernest Henry Starling," http://www.whonamedit.com/doctor.cfm/1188.html.

Expert Panel on Detection, Evaluation, and Treatment of High Blood Cholesterol in Adults. "Executive Summary of the Third Report of the National Cholesterol Education Program (NCEP) Expert Panel on Detection, Evaluation, and Treatment of High Blood Cholesterol in Adults (Adult Treatment Panel III)." *Journal of the American Medical Association* 285 (16 May 2001): 2486–2497.

Friedman, Meyer, and Gerald W. Friedland. *Medicine's 10 Greatest Discoveries.* New Haven, CT: Yale University Press, 1998.

Friend, Tim. "Researchers Learning to End Tumor's Blood Supply." *USA Today*, 15 May 2002. http://www.usatoday.com/news/health/cancer/2002-05-16-angiogenesis.htm.

"Forgotten Transfusion History: John Leacock of Barbados." http://www.pubmed central.nih.gov/articlerender.fcgi?artid=139049.

Gorman, Christine. "Repairing the Damage." *Time* 157, no. 5 (5 February 2001): 52–58.

Harder, Ben. "Vitamin Void: Heart Disease May Lurk in B12 Deficiency." *Science News* 161, no. 7 (16 February 2002): 100.

———. "Wholesome Grains: Insulin Effects May Explain Healthful Diet." *Science News* 161, no. 20 (18 May 2002): 308.

"The Heart and Circulatory System." http://web.ukonline.co.uk/webwise/spinneret/circuln/heart.htm.

"The Heart: An Online Exploration." http://sln.fi.edu/biosci/heart.html.

"Heart Laser Surgery: An Alternative to Transplantation." http://www.newswise.com/articles/view/7094/.

Heeschen, C., J. J. Jang, M. Weis, A. Pathak, S. Kaji, R. S. Hu, P. S. Tsao, F. L. Johnson, and J. P. Cooke. "Nicotine Stimulates Angiogenesis and Promotes Tumor Growth and Atherosclerosis." *Nature Medicine* 7 (1 July 2001): 833–839.

"Heinrich Müller." http://www.whonamedit.com/doctor.cfm/2564.html.

"Helen Brooke Taussig." http://www.whonamedit.com/doctor.cfm/2034.html.

"High Blood Cholesterol: What You Need to Know." http://www.nhlbi.nih.gov/health/public/heart/chol/wyntk.htm.

"Highlights to Transfusion Medicine History 1900 to 1947." http://www.blood services.org/bloodlines/vol_3_no_11/06_history02.htm.

"The History of Blood Transfusion Medicine." http://www.bloodbook.com/trans-history.html.

"HIV Infection and AIDS: An Overview." http://www.aidsinfo.nih.gov/ed_resources/default.asp?cat_id=P.

Hsue, Priscilla Y., Cynthia L. Salinas, Ann F. Bolger, Neal L. Benowitz, and David D. Waters. "Acute Aortic Dissection Related to Crack Cocaine." *Circulation* 105 (April 2002): 1592–1595.

Hurst, J. Willis, and Fye, W. Bruce, eds. "Rudolf Albert von Koelliker." http://www.clinicalcardiology.org/briefs/9905briefs/22-376.html.

Institute of Medicine of National Academics. "Physical Activity." In *Dietary Reference Intakes for Energy, Carbohydrate, Fiber, Fat, Fatty Acids, Cholesterol, Protein, and Amino Acids (Macronutrients)*. Washington, D.C.: 697–736. National Academies Press, 2002.

"Karl Landsteiner and Alexander Wiener." http://www.bloodmed.com/home/hannpdf/bjh2139.pdf.

"Karl Landsteiner—Discoverer of the Major Human Blood Groups." http://www.
        mayo.edu/proceedings/2001/aug/7608sv.pdf.

Kaufman, Dan S., Eric T. Hanson, Rachel L. Lewis, Robert Auerbach, and James A.
        Thomson. "Hematopoietic Colony-Forming Cells Derived from Human Em-
        bryonic Stem Cells." *Proceedings of the National Academy of Sciences* 98, no.
        19 (11 September 2001): 10716–10721.

Kenchaiah, Satish, Jane C. Evans, Daniel Levy, Peter W.F. Wilson, Emelia J. Ben-
        jamin, Martin G. Larson, William B. Kannel, and Ramachandran S. Vasan.
        "Obesity and the Risk of Heart Failure." *New England Journal of Medicine* 347,
        no. 5 (1 August 2002): 305–313.

Klabunde, Richard E. "Cardiovascular Physiology Concepts." http://www.cvphysi
        ology.com.

Lang, Leslie H. "Platelet Molecule Regulates Blood Coagulation, Study Finds." 25
        July 2002. http://www.unc.edu/news/newsserv/archives/jul02/thromb072402
        .htm.

Levick, J. R. *An Introduction to Cardiovascular Physiology*, 2nd ed. Oxford:
        Butterworth-Heinemann, 1995.

"Lifeblood." http://sln.fi.edu/biosci/blood/blood.html.

Loudon, Irvine, ed. *Western Medicine: An Illustrated History.* Oxford: Oxford Uni-
        versity Press, 1997.

"Luigi   Galvani   (1737–1798)."   http://www.corrosion-doctors.org/Biographies/
        GalvaniBio.htm.

Lyons, Albert S., and R. Joseph Petrucelli II. *Medicine: An Illustrated History.* New
        York: Aberdale Press, 1987.

Margolis, Simeon, ed. *The Johns Hopkins Medical Handbook: The 100 Major Med-
        ical Disorders of People Over the Age of 50.* New York: Rebus, Inc., 1999.

"Maude   Elizabeth   Seymour   Abbott."   http://www.whonamedit.com/doctor.cfm/
        113.html.

"Max F. Perutz—Biography." http://www.nobel.se/chemistry/laureates/1962/perutz-
        bio.html.

Mertz, Leslie. "The Science of Physiology: Galen's Influence." In *Science and Soci-
        ety through Time*, ed. Neil Schlager. Detroit: Gale, 2000, 125–128.

Meyers, D.G., D. Strickland, P.A. Maloley, J.K. Seburg, J.E. Wilson, and B.F. Mc-
        Manus. "Possible Association of a Reduction in Cardiovascular Events with
        Blood Donation." *Heart* 78 (August 1997): 188–193.

Miller, D.G., and G. Stamatoyannopoulos. "Gene Therapy for Hemophilia." *New En-
        gland Journal of Medicine* 344 (June 2001): 1782–1784.

Miller, Karen, and Tony Phillips. "Patches for a Broken Heart." NASA. http://
        science.nasa.gov/headlines/y2002/14feb_heart.htm?list154233.

Mittleman, Murray A., Rebecca A. Lewis, Malcolm Maclure, Jane B. Sherwood, and
        James E. Muller. "Triggering Myocardial Infarction by Marijuana." *Circulation*
        103 (June 2001): 2805–2809.

Mittleman, Murray A., David Mintzer, Malcolm Maclure, Geoffrey H. Tofler, Jane B.
        Sherwood, and James E. Muller. "Triggering of Myocardial Infarction by Co-
        caine." *Circulation* 99 (June 1999): 2737–2741.

Moolgavkar, Suresh H. "Air Pollution and Hospital Admissions for Diseases of the
        Circulatory System in Three U.S. Metropolitan Areas." *Journal of the Air &
        Waste Management Association* 50, no. 7 (July 2000): 1199–1206.

Mukamal, Kenneth J. "Roles of Drinking Pattern and Type of Alcohol Consumed in
        Coronary Heart Disease in Men." *New England Journal of Medicine* 348, no. 2
        (9 January 2003): 109–118.

"Myocardial Adaptations to Training." http://home.hia.no/~stephens/hrttrn.htm.

Netting, Jessa. "Telltale Heart: Researchers Are Uncovering the Genetic Plan for Building a Heart." *Science News* 160 (7 July 2001): 13.

Nienaber C. A., R. Fattori, G. Lund, C. Dieckmann, W. Wolf, Y. von Kodolitsch, V. Nicolas, and A. Pierangeli. "Nonsurgical Reconstruction of Thoracic Aortic Dissection by Stent–Graft Placement." *New England Journal of Medicine* 340 (20 May 1999): 1539–1545.

Persaud, T. V. N. *Early History of Human Anatomy: From Antiquity to the Beginning of the Modern Era.* Springfield, IL: Charles C. Thomas, 1984.

Piña, Ileana L., Carl S. Apstein, Gary J. Balady, Romualdo Belardinelli, Bernard R. Chaitman, Brian D. Duscha, Barbara J. Fletcher, Jerome L. Fleg, Jonathan N. Myers, and Martin J. Sullivan. "Exercise and Heart Failure: A Statement from the American Heart Association Committee on Exercise, Rehabilitation, and Prevention." *Circulation* 107 (March 2003): 1210–1225.

Pradhan, Aruna D., JoAnn E. Manson, Nader Rifai, Julie E. Buring, and Paul M. Ridker. "C-Reactive Protein, Interleukin 6, and Risk of Developing Type 2 Diabetes Mellitus." *Journal of the American Medical Association* 286 (18 July 2001): 327–334.

Raloff, Janet. "Blood Points to Pollutions Heart Risks." *Science News* 160 (7 July 2001): 9.

Rappaport, Samuel, and Helen Wright. *Great Adventures in Medicine.* New York: Dial Press, 1952.

Raz, Itamar, Dana Elias, Ann Avron, Merana Tamir, Muriel Metzger, and Irun R. Cohen. "ß-cell Function in New-Onset Type 1 Diabetes and Immunomodulation with a Heat-Shock Protein Peptide (DiaPep277): A Randomised, Double-Blind, Phase II Trial." *Lancet* 358 (24 November 2001): 1749.

"Red Gold: The Epic Story of Blood." http://www.pbs.org/wnet/redgold/journey/phase2_a1.html.

"Rh Disease." http://www.marchofdimes.com/professionals/681_1220.asp.

Sacks, Frank M., Walter C. Willett, Angela Smith, Lisa Brown, Bernard Rosner, and Thomas Moore. "Effect on Blood Pressure of Potassium, Calcium, and Magnesium in Women with Low Habitual Intake." *Hypertension* 31 (January 1998): 131–138.

Seppa, N. "Drugs Slow Diabetes' Patients Kidney Damage." *Science News* 160 (22 September 2001): 182.

———. "Gene Therapy for Sickle-Cell Disease?" *Science News* 160 (15 December 2001): 372.

———. "Thinking Blurs when Blood Sugar Strays." *Science News* 160 (21 July 2001): 47.

———. "Two Drugs May Enhance Recovery from Stroke." *Science News* 160 (15 September 2001): 166.

Shaw, M., Senior ed. *Everything You Need to Know about Disease.* Springhouse, PA: Springhouse Corp, 1996.

Shephard, Roy J., and Gary J. Balady. "Exercise as Cardiovascular Therapy." *Circulation* 99 (February 1999): 963–972.

Simon, Seymour. *The Heart: Our Circulatory System.* New York: Morrow Junior Books, 1996.

Singer, W., T. L. Opfer-Gehrking, B. R. McPhee, M. J. Hilz, A. E. Bharucha, and P. A. Low. "Acetylcholinesterase Inhibition: A Novel Approach in the Treatment of Neurogenic Orthostatic Hypotension." *Journal of Neurology and Neurosurgical Psychiatry* 74 (September 2003): 1294–1298.

Stefanadis, Christodoulos, Eleftherios Tsiamis, Charalambos Vlachopoulos, Costas Stratos, Konstantinos Toutouzas, Christos Pitsavos, Stelios Marakas, Harisios Boudoulas, and Pavlos Toutouzas. "Unfavorable Effect of Smoking on the Elastic Properties of the Human Aorta." *Circulation* 95 (January 1997): 31–38.

Thibodeau, Gary A. *Anthony's Book of Anatomy and Physiology,* 13th ed. New York: Times Mirror/Mosby College Publishing, 1990.

"Timeline: A Brief History of AIDS/HIV." http://www.aegis.com/topics/timeline/default.asp.

Travis, John. "Stem Cell Research Marches On." *Science News* 160 (1 September 2001): 143.

"'Trojan Horse' Technology Destroys Blood Supply to Cancer Tumors in Mice." 11 June 2002. http://irweb.swmed.edu/newspub/newsdetl.asp?story_id=424.

"UM Researchers Discover 'Key' to Blood-Brain Barrier." http://www.umm.edu/news/releases/bloodbrain.html.

Vasan, Ramachandran S., Martin G. Larson, Eric P. Leip, Jane C. Evans, Christopher J. O'Donnell, William B. Kannel, and Daniel Levy. "Impact of High-Normal Blood Pressure on the Risk of Cardiovascular Disease." *New England Journal of Medicine* 345 (1 November 2001): 1291–1297.

Vogel, Steven. *Vital Circuits: On Pumps, Pipes, and the Workings of Circulatory Systems.* New York: Oxford University Press, 1992.

Wang, L. "Blood Relatives: First-Generation Artificial Blood Is About to Hit the Market." *Science News* 159, no. 13 (31 March 2001): 206–207.

Weisse, Allen B. *Medical Odysseys: The Different and Sometimes Unexpected Pathways to 20th Century Medical Discoveries.* New Brunswick, NJ: Rutgers University Press, 1991.

"Werner Forssmann (1904–1979)." http://www.ptca.org/archive/bios/forssmann.html.

"Werner Forssmann—Biography." http://www.nobel.se/medicine/laureates/1956/forssmann-bio.html.

Whelton, Paul, Jiang He, Lawrence J. Appel, Jeffrey A. Cutler, Stephen Havas, Theodore A. Kotchen, Edward J. Roccella, Ron Stout, Carlos Vallbona, Mary C. Winston, and Joanne Karimbakas, for the National High Blood Pressure Education Program Coordinating Committee. "Primary Prevention of Hypertension: Clinical and Public Health Advisory from the National High Blood Pressure Education Program." *The Journal of the American Medical Association* 288 (16 October 2002): 1882–1888.

Whitfield, Philip, general ed. *The Human Body Explained: A Guide to Understanding the Incredible Living Machine.* New York: Henry Holt and Co., 1995.

Writing Group of the PREMIER Collaborative Research Group. "Effects of Comprehensive Lifestyle Modification on Blood Pressure Control: Main Results of the PREMIER Clinical Trial." *Journal of the American Medical Association* 289 (23 April 2003): 2083–2093.

Zipursky, A. "The Rise and Fall of Rh Disease." In *Historical Review and Recent Advances in Neonatal and Perinatal Medicine.* Evansville, IN: Mead Johnson Nutritional Division, 1980. http://www.neonatology.org/classics/mj1980/ch07.html.

Index

About the Author

LESLIE MERTZ is a biologist at Wayne State University in Detroit, Michigan. She is the author of *Recent Advances and Issues in Biology* (Oryx Press, 2000), and former editor of the research magazine *New Science*.